RIGHT-ANGLED TRIANGLES

NCEA Level 1 Internal

Charlotte Walker and Victoria Walker

Walker Maths 1.7 Right-angled triangles
1st Edition
Charlotte Walker
Victoria Walker

Editor: Eva Chan
Designer: Cheryl Smith, Macarn Design
Production controller: Siew Han Ong

Any URLs contained in this publication were checked for currency during the production process. Note, however, that the publisher cannot vouch for the ongoing currency of URLs.

Acknowledgements
Cover photo courtesy of Shutterstock.

We wish to thank the Boards of Trustees of Darfield and Riccarton High Schools for allowing us to use materials and ideas developed while teaching. Our thanks also go to all past and present colleagues who have generously shared their expertise and ideas.

For product information and technology assistance,
in Australia call **1300 790 853**;
in New Zealand call **0800 449 725**

For permission to use material from this text or product, please email **aust.permissions@cengage.com**

National Library of New Zealand Cataloguing-in-Publication Data
A catalogue record for this book is available from the National Library of New Zealand.

978 0 17 037162 9

Cengage Learning Australia
Level 7, 80 Dorcas Street
South Melbourne, Victoria, Australia 3205

Cengage Learning New Zealand
Unit 4B Rosedale Office Park
331 Rosedale Road, Albany, North Shore 0632, NZ

For learning solutions, visit **cengage.co.nz**

Printed in China by 1010 Printing International Limited
12 13 14 15 25 24 23

CONTENTS

Formulae

These are the formulae for this achievement standard. Remember to check with your teacher to see which ones you will be provided with in your assessment.

Theorem of Pythagoras	$a^2 + b^2 = c^2$
Trigonometry	O / S H $\sin\theta = \frac{O}{H}$ A / C H $\cos\theta = \frac{A}{H}$ O / T A $\tan\theta = \frac{O}{A}$

Glossary

Make your own glossary of key terms:

Term	Definition	Picture/Example
Theorem of Pythagoras		
Hypotenuse		
Opposite		
Adjacent		

ISBN: 9780170371629

Term	Definition	Picture/Example
Equilateral triangle		
Isosceles triangle		
Scalene triangle		
Sine (sin)		
Cosine (cos)		
Tangent (tan)		
Similar shapes		
Horizontal		
Vertical		
Theta (θ)		

ISBN: 9780170371629

Metric units

Length

Lengths and distances are usually measured in mm, cm, m or km.

Converting units

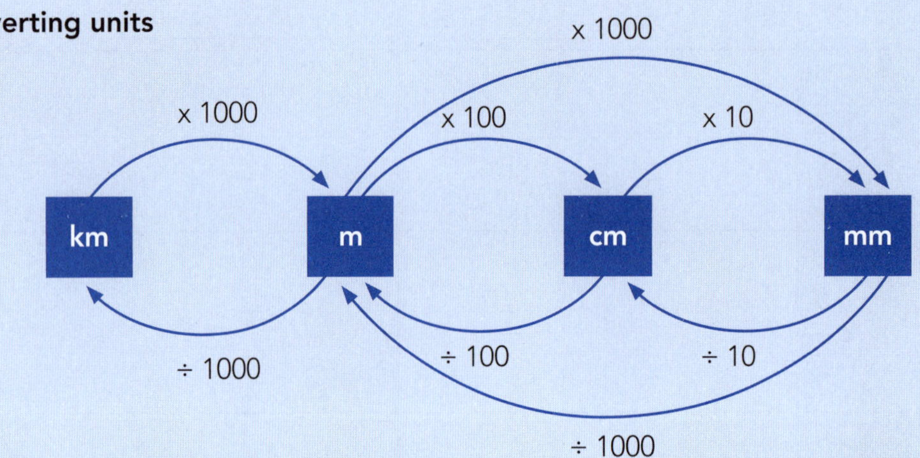

Example one: Convert 0.65 km to m.

1 km = 1000 m
∴ 0.65 km = 0.65 **x** 1000 m
= 650 m

Your answer will be a bigger number, so **multiply**.

Example two: Convert 1275 mm to m.

1000 mm = 1 m
∴ 1275 mm = 1275 **÷** 1000 m
= 1.275 m

Your answer will be a smaller number, so **divide**.

Convert the following.

1	27 cm = ______________ mm	**2**	8 m = ______________ cm
3	2 km = ______________ m	**4**	9 cm = ______________ mm
5	95 mm = ______________ cm	**6**	1400 m = ______________ km
7	17 m = ______________ cm	**8**	82 mm = ______________ cm
9	1700 mm = ______________ cm	**10**	11.3 cm = ______________ mm
11	685 cm = ______________ m	**12**	2.38 km = ______________ m
13	0.7 km = ______________ m	**14**	372 cm = ______________ m
15	2.64 km = ______________ m	**16**	8.15 m = ______________ cm
17	7350 m = ______________ km	**18**	5.6 m = ______________ cm
19	0.04 km = ______________ m	**20**	6 cm = ______________ m

ISBN: 9780170371629

Area

- Length is measured in mm, cm, m or km.
- Area is measured in mm^2, cm^2, m^2, ha (hectares) or km^2.
- When converting between these, be very careful.

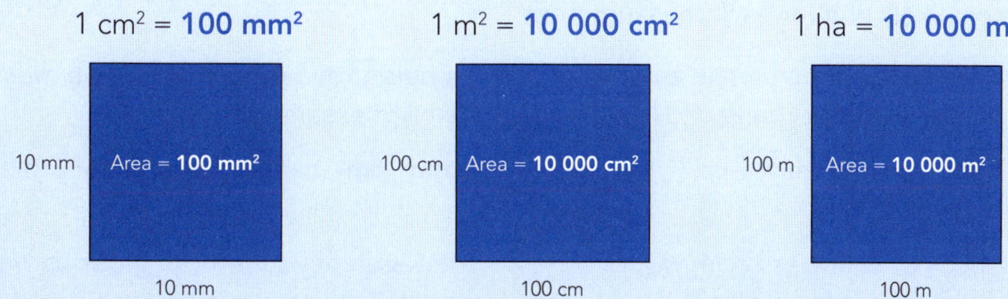

Example one: Convert 14.05 cm^2 into mm^2.

$$1 \text{ cm}^2 = 100 \text{ mm}^2$$
$$\therefore 14.05 \text{ cm}^2 = 100 \times 14.05 \text{ mm}^2$$
$$= 1405 \text{ mm}^2$$

Your answer will be a bigger number, so **multiply**.

Example two: A tennis court is 11 m wide and 24 m long. Calculate its area in hectares.

$$\text{Area} = 11 \text{ m} \times 24 \text{ m}$$
$$= 264 \text{ m}^2$$
$$10\,000 \text{ m}^2 = 1 \text{ ha}$$
$$\therefore 264 \text{ m}^2 = \frac{264}{10\,000} \text{ ha}$$
$$= 0.0264 \text{ ha}$$

Your answer will be a smaller number, so **divide**.

Convert the following areas.

1 2 m^2 = ____________ cm^2

2 170 mm^2 = ____________ cm^2

3 4 cm^2 = ____________ mm^2

4 1546 mm^2 = ____________ cm^2

5 1.2 m^2 = ____________ cm^2

6 6932 m^2 = ____________ ha

7 2.56 ha = ____________ m^2

8 0.08 ha = ____________ m^2

9 0.604 ha = ____________ m^2

10 12 750 m^2 = ____________ ha

11 765 mm^2 = ____________ cm^2

12 1500 m^2 = ____________ ha

13 A rug measures 180 cm by 238 cm. Calculate its area in m^2.

____

____

14 A rectangular section has an area of 0.144 ha. If it is 36 m long, how wide is it?

____

____

ISBN: 9780170371629

Sensible units

- Length is measured in mm, cm, m or km.
- Area is measured in mm^2, cm^2, m^2, ha (hectares) or km^2.

Circle the most likely unit for each measurement.

1 The height of an ice cream in a cone would most likely be measured in:

mm cm m km mm^2 cm^2 m^2 ha

2 A height of a person would be most likely measured in:

mm cm m km mm^2 cm^2 m^2 ha

3 The length of your classroom would most likely be measured in:

mm cm m km mm^2 cm^2 m^2 ha

4 The area of your school grounds would most likely be measured in:

mm cm m km mm^2 cm^2 m^2 ha

5 The area of your fingernail would most likely be measured in:

mm cm m km mm^2 cm^2 m^2 ha

6 The distance from your home to school would most likely be measured in:

mm cm m km mm^2 cm^2 m^2 ha

7 The area of your bedroom would most likely be measured in:

mm cm m km mm^2 cm^2 m^2 ha

8 The height your school gymnasium would most likely be measured in:

mm cm m km mm^2 cm^2 m^2 ha

9 The width of your nose would most likely be measured in:

mm cm m km mm^2 cm^2 m^2 ha

10 The area of a plate would most likely be measured in:

mm cm m km mm^2 cm^2 m^2 ha

11 The length of a beetle would most likely be measured in:

mm cm m km mm^2 cm^2 m^2 ha

12 The width of this book would most likely be measured in:

mm cm m km mm^2 cm^2 m^2 ha

13 The distance a discus is thrown would most likely be measured in:

mm cm m km mm^2 cm^2 m^2 ha

14 The width of a piano key would most likely be measured in:

mm cm m km mm^2 cm^2 m^2 ha

15 The length of your gut would most likely be measured in:

mm cm m km mm^2 cm^2 m^2 ha

16 The surface area of your skin would most likely be measured in:

mm cm m km mm^2 cm^2 m^2 ha

17 The surface area of a leaf would most likely be measured in:

mm cm m km mm^2 cm^2 m^2 ha

18 The area of a forest would most likely be measured in:

mm cm m km mm^2 cm^2 m^2 ha

ISBN: 9780170371629

Estimating lengths and areas

Find the most likely quantities from the column on the right to complete the sentences on the left.

1 The length of your calculator would most likely be ____________.	**2 mm**
2 The circumference of your head is about ____________.	**8 m**
3 The average length of the South Island is about ____________.	**0.5 ha**
4 The diameter of a pencil lead would be about ____________.	**55 cm**
5 A length of your classroom would most likely be ____________.	**9 m^2**
6 The length of a rugby field is about ____________.	**0.05 ha**
7 The area of a rugby field is about ____________.	**30 m**
8 The area of the front cover of this book is about ____________. (Don't measure it!)	**3.5 m^2**
9 The area of your bedroom would most likely be ____________.	**15 cm**
10 The length of your school hall would most likely be ____________.	**120 cm^2**
11 The surface area of your skin would most likely be ____________.	**100 m**
12 The average height of a Year 9 student would be about ____________.	**8 ha**
13 The surface area of a potato would be about ____________.	**840 km**
14 The area of your school grounds would most likely be ____________.	**1.5 m^2**
15 The area of your section at home would most likely be ____________.	**600 cm^2**
16 The area of the whiteboard in your classroom would most likely be ____________.	**1.5 m**

ISBN: 9780170371629

Theorem of Pythagoras

- The Theorem of Pythagoras applies to **right-angled** triangles only.
- The longest side is the **hypotenuse**.
- The hypotenuse is always **opposite** the right angle.
- The theorem is used for finding **lengths** of sides.

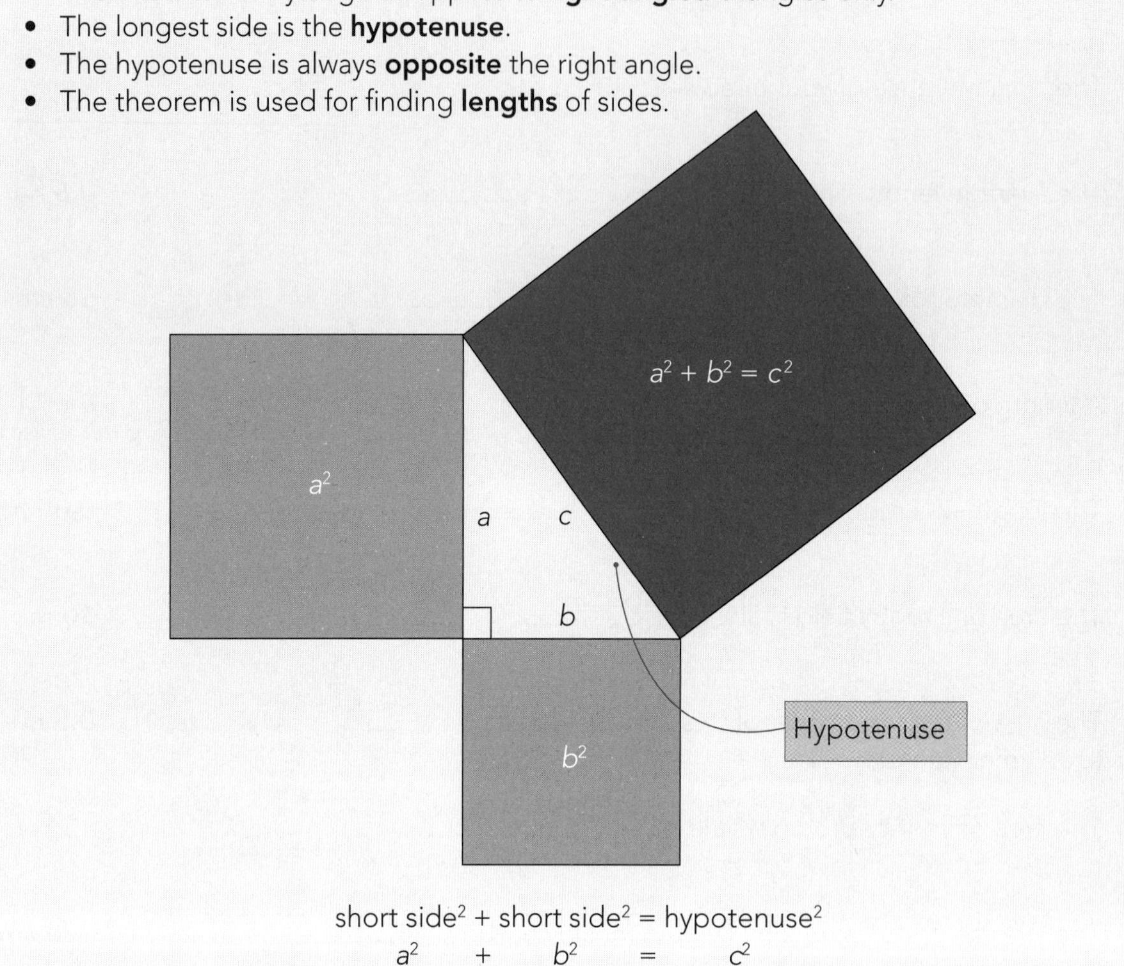

short side^2 + short side^2 = hypotenuse^2

$$a^2 + b^2 = c^2$$

Finding the length of the hypotenuse

Example: Calculate the length of *c*.

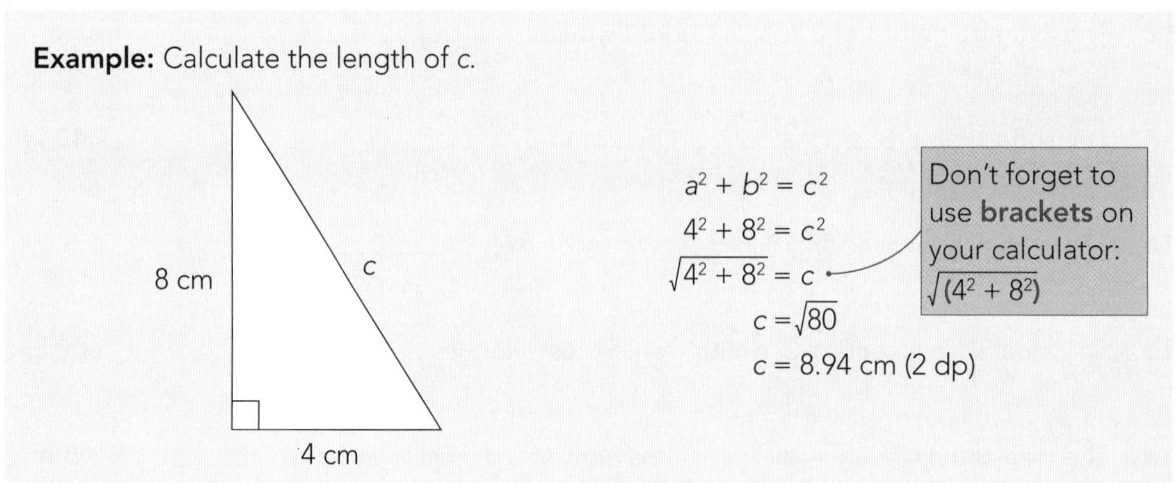

$$a^2 + b^2 = c^2$$
$$4^2 + 8^2 = c^2$$
$$\sqrt{4^2 + 8^2} = c$$
$$c = \sqrt{80}$$
$$c = 8.94 \text{ cm (2 dp)}$$

  ISBN: 9780170371629

Calculate the unknown lengths of each triangle.

1

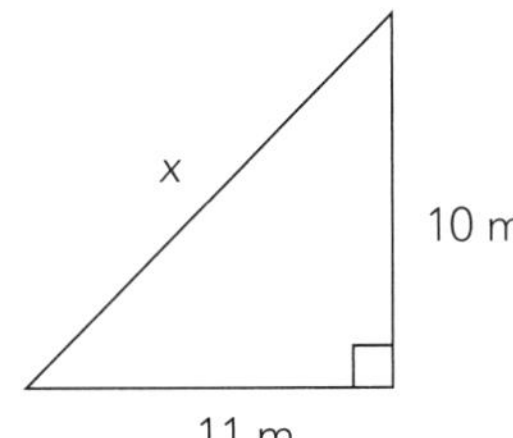

2

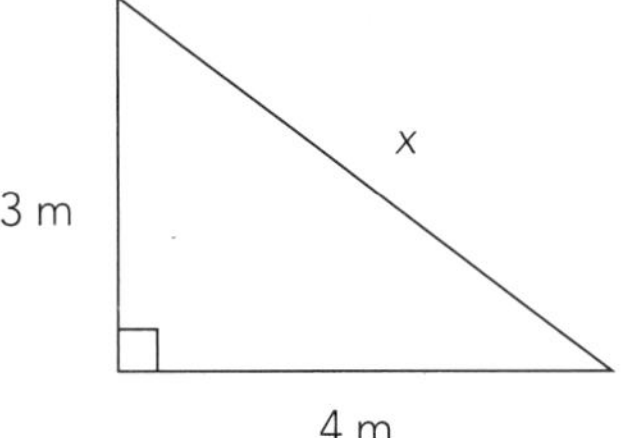

3

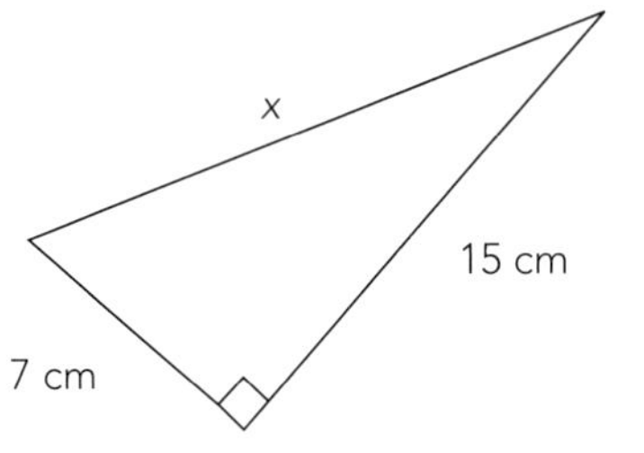

4

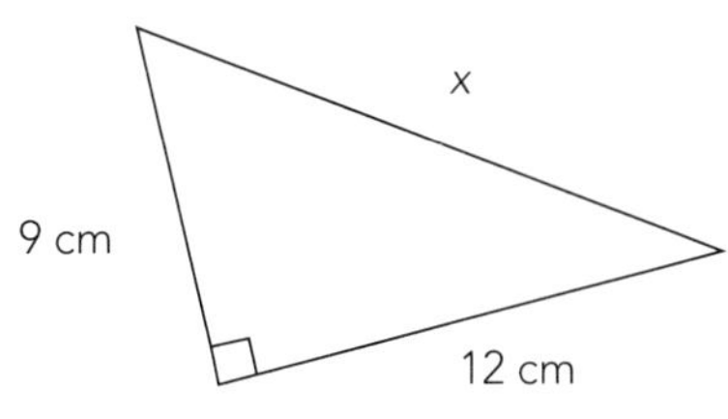

5

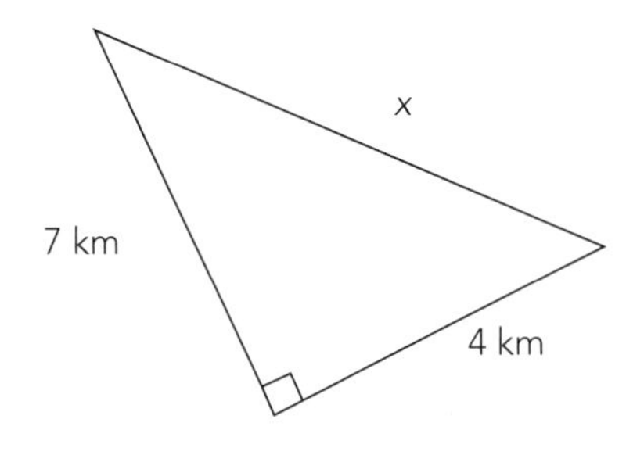

6

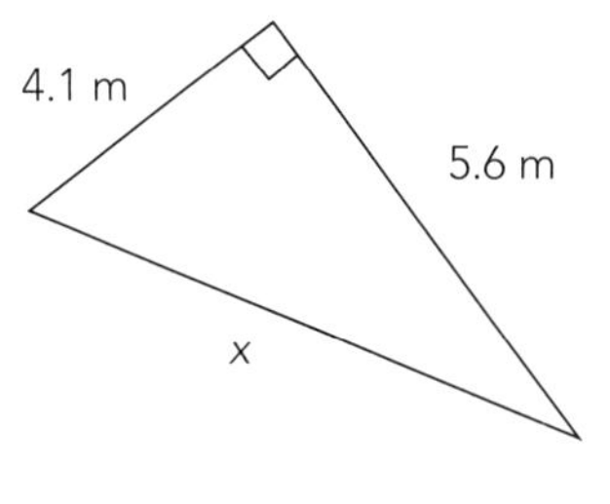

7

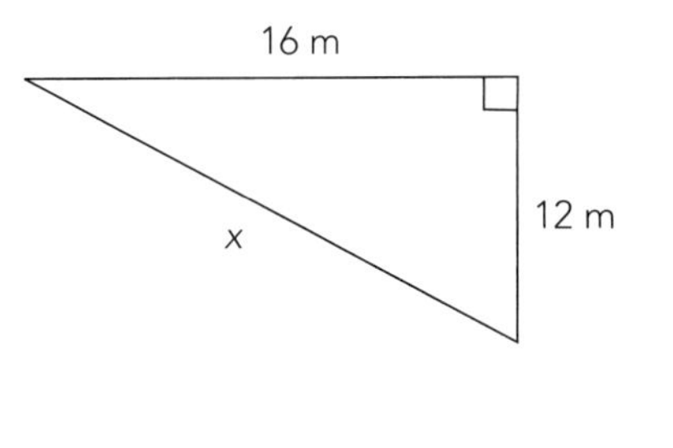

8

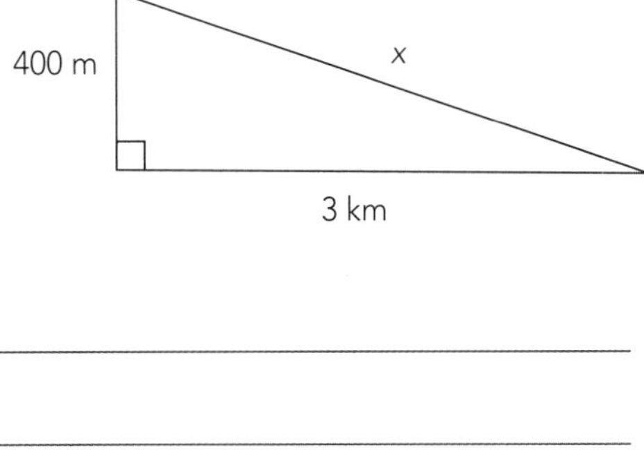

ISBN: 9780170371629

9

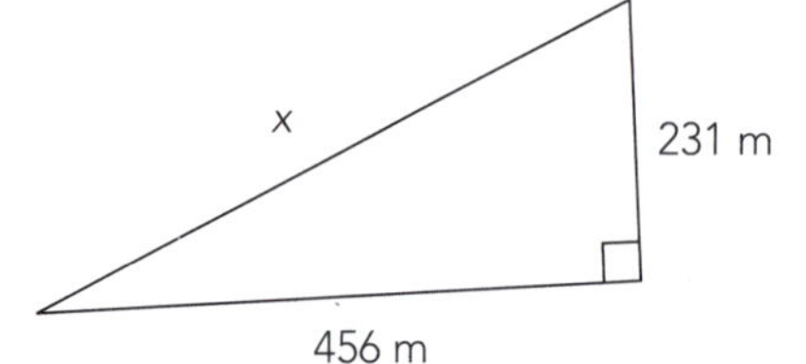

10

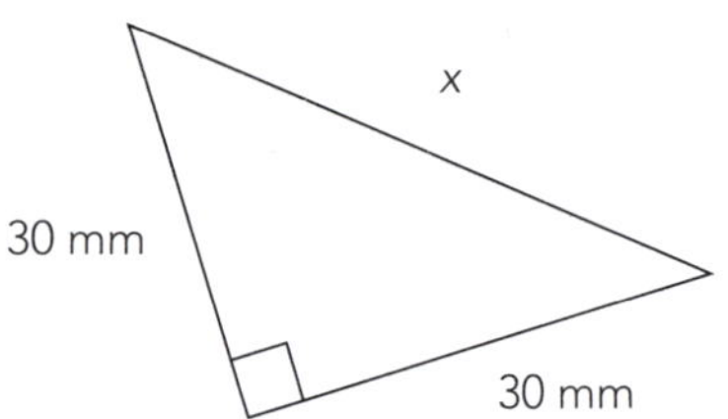

11

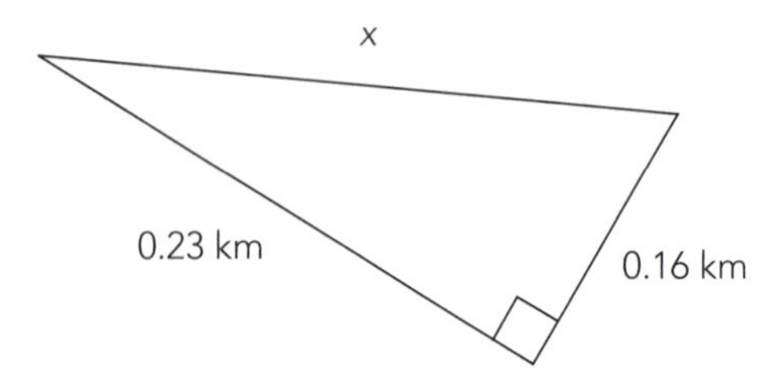

12

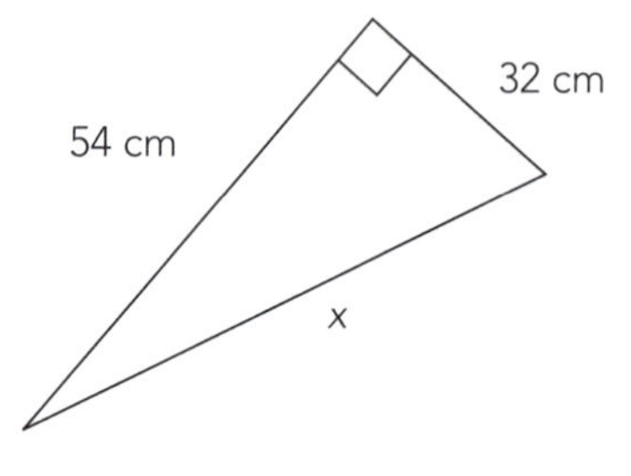

13

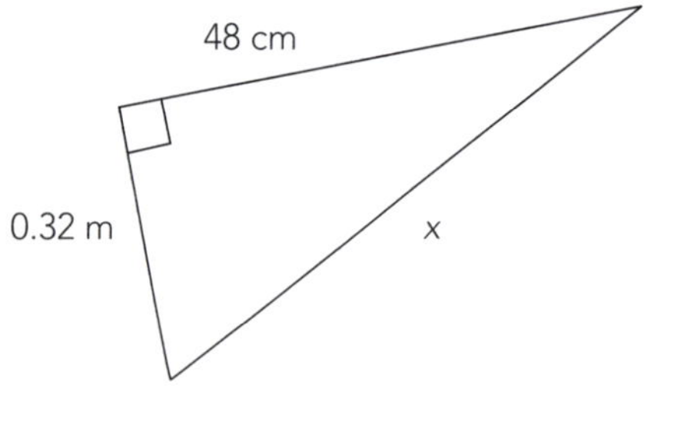

14

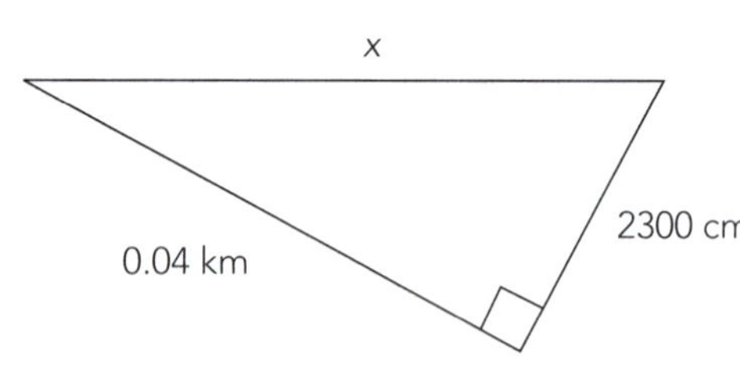

15

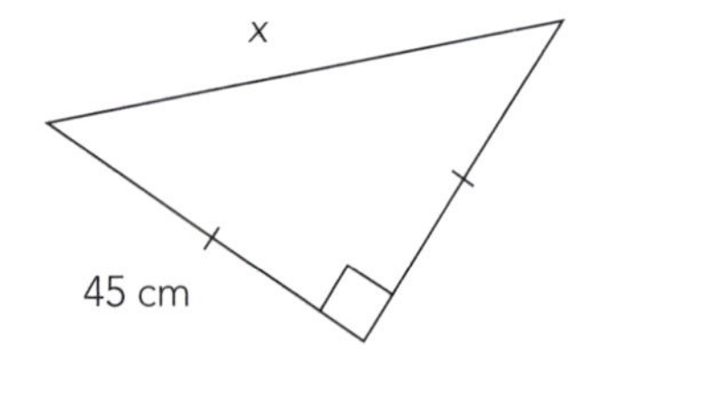

16

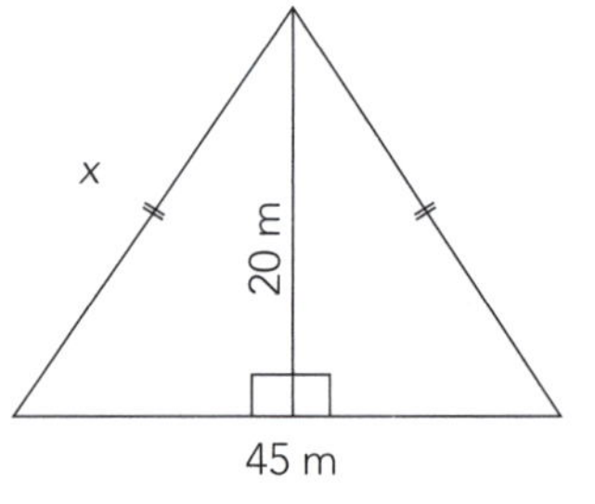

ISBN: 9780170371629

Finding the lengths of short sides

Example: Calculate the length of y.

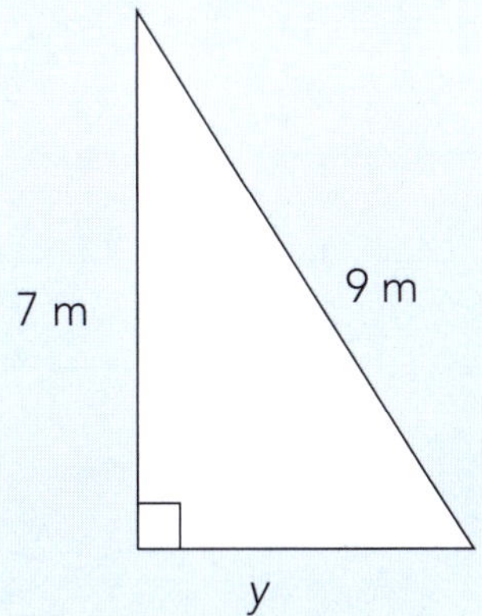

$a^2 + b^2 = c^2$

$y^2 + 7^2 = 9^2$

$y^2 = 9^2 - 7^2$

$y = \sqrt{9^2 - 7^2}$

$y = \sqrt{32}$

$y = 5.66$ m (2 dp)

Don't forget to use **brackets** on your calculator: $\sqrt{(9^2 - 7^2)}$

Calculate the unknown lengths of each triangle.

1

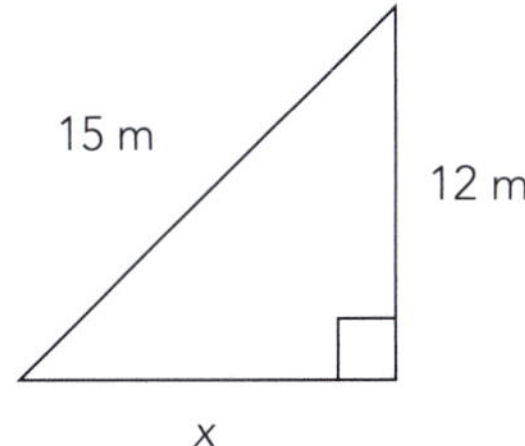

2

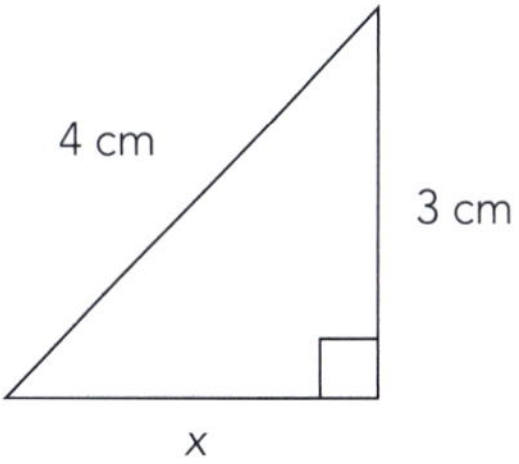

3

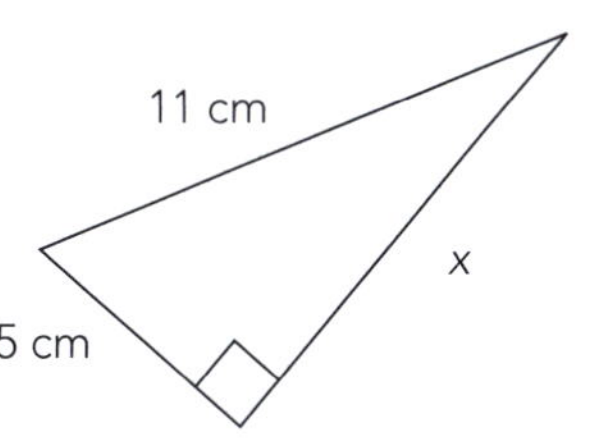

4

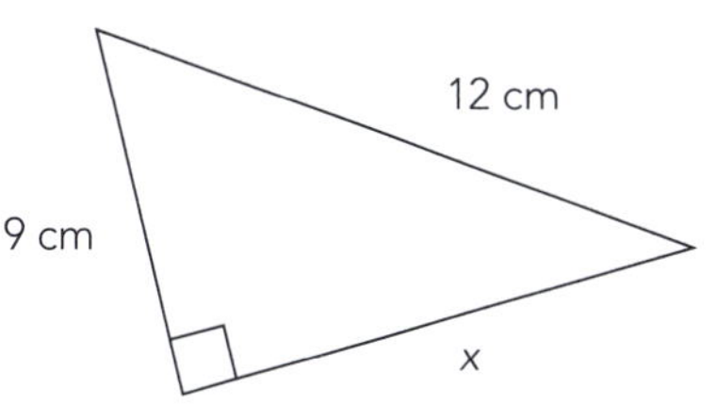

5

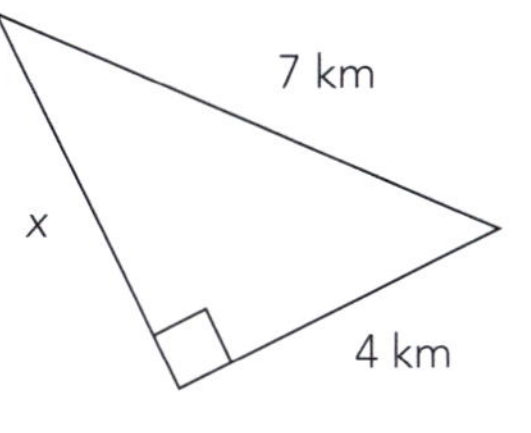

6

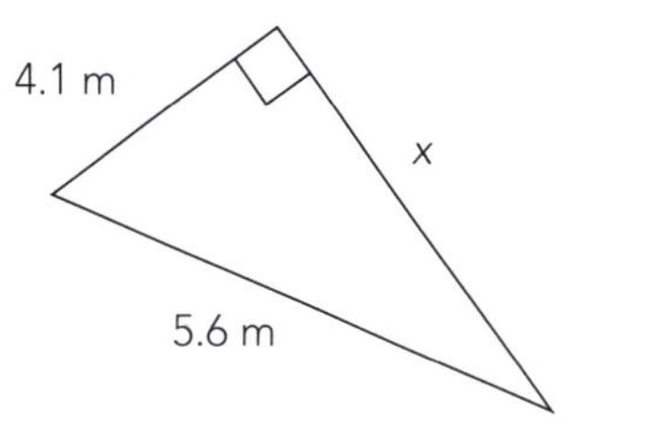

ISBN: 9780170371629

7

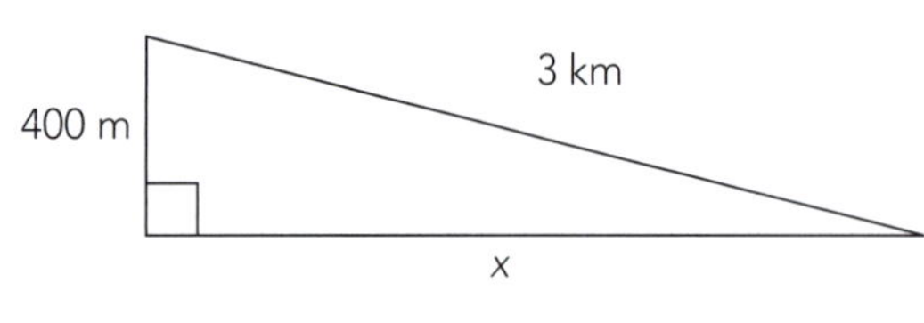

8

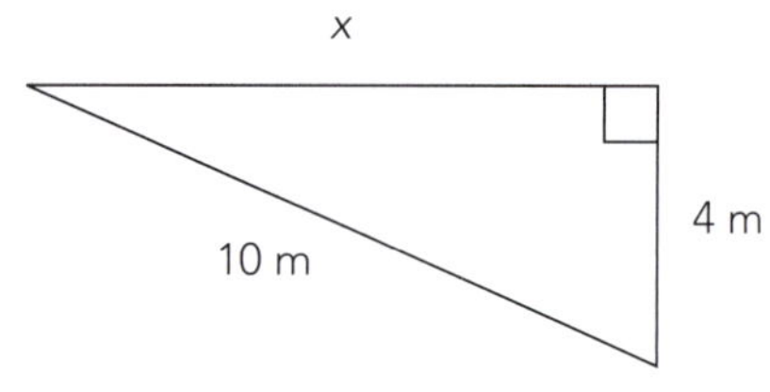

9

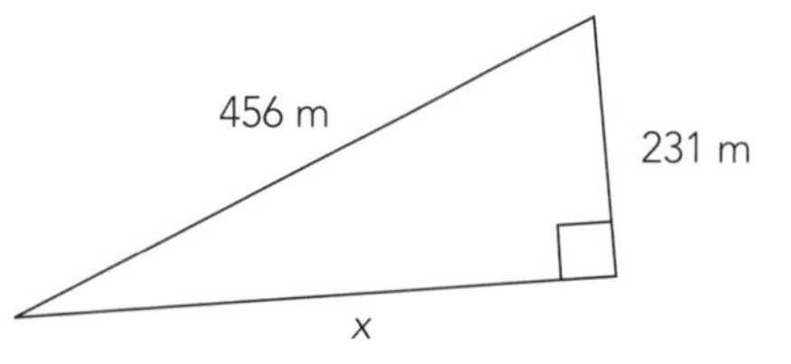

10

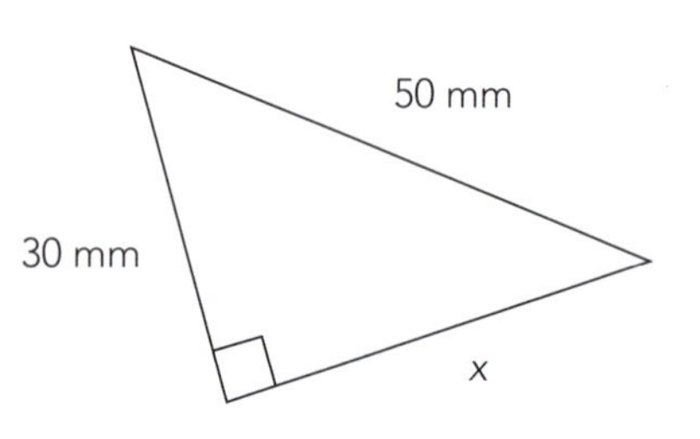

11

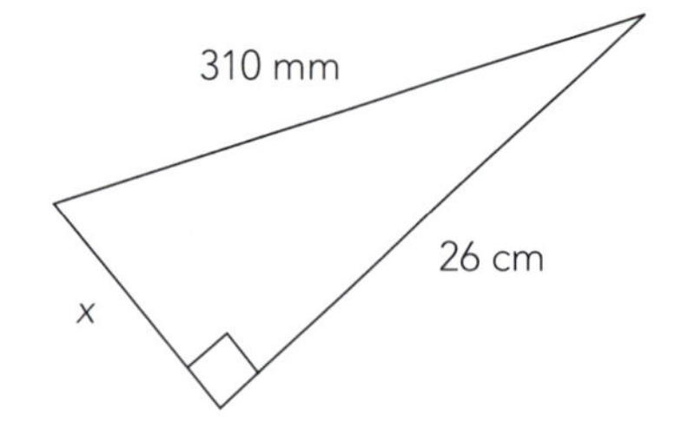

12

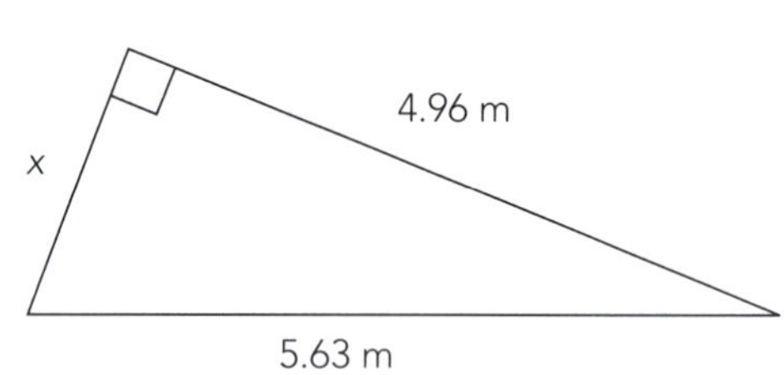

13

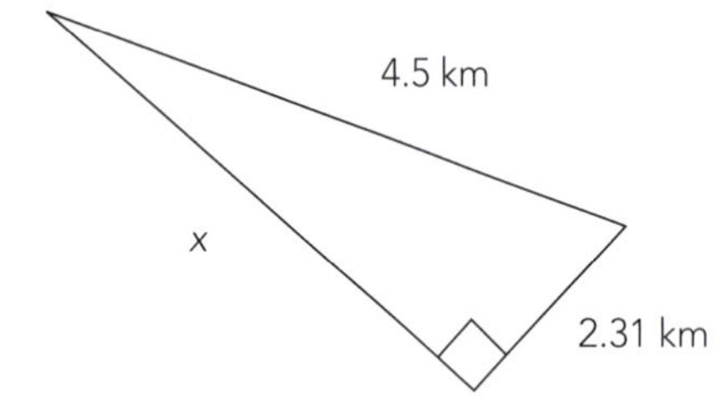

14 The diagram shows an equilateral triangle that has been split into two smaller triangles. Calculate the height (*h*).

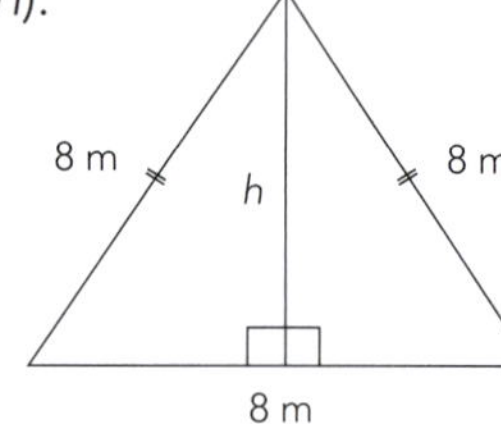

 ISBN: 9780170371629

Mixing it up

Find the missing sides of these triangles.

1

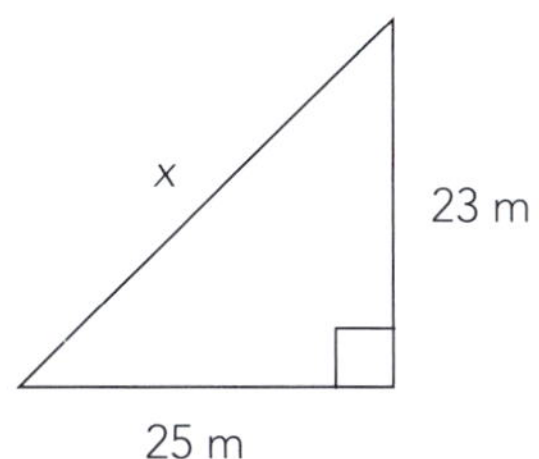

2

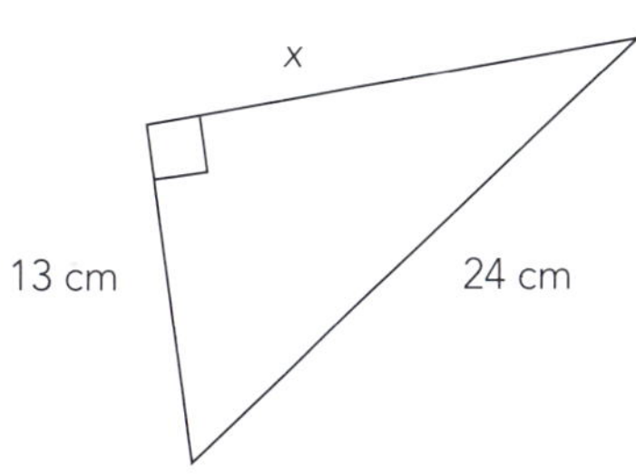

3

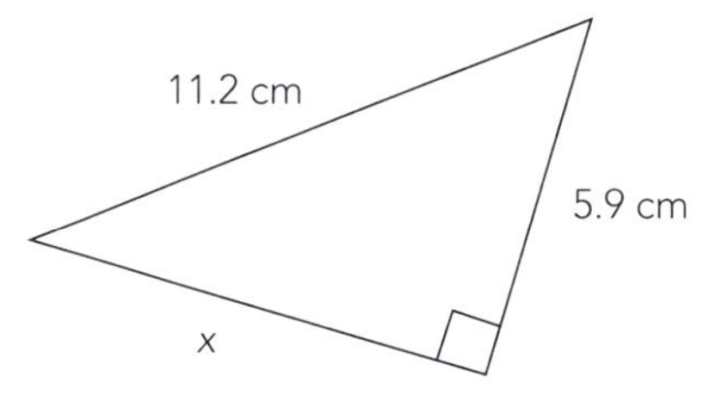

4

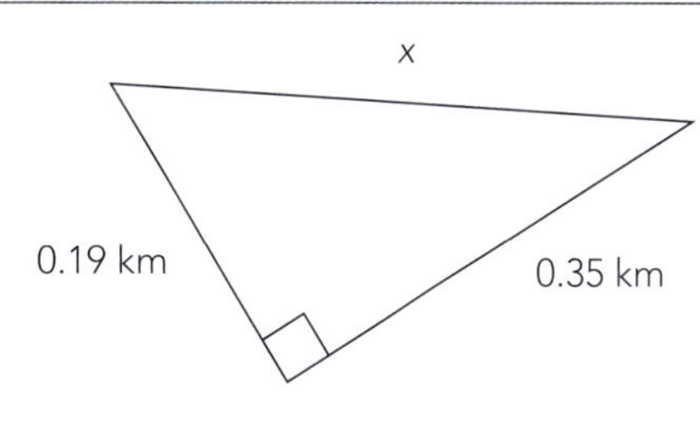

5

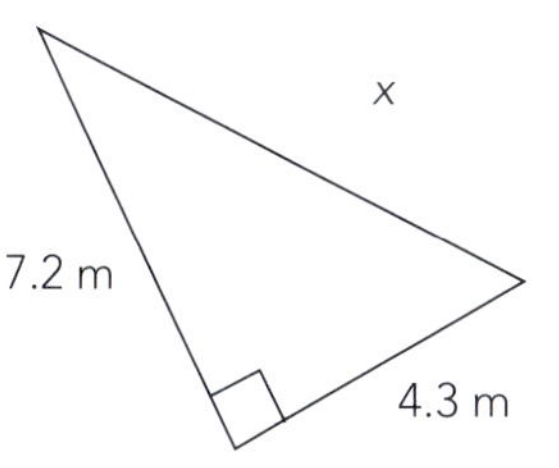

6

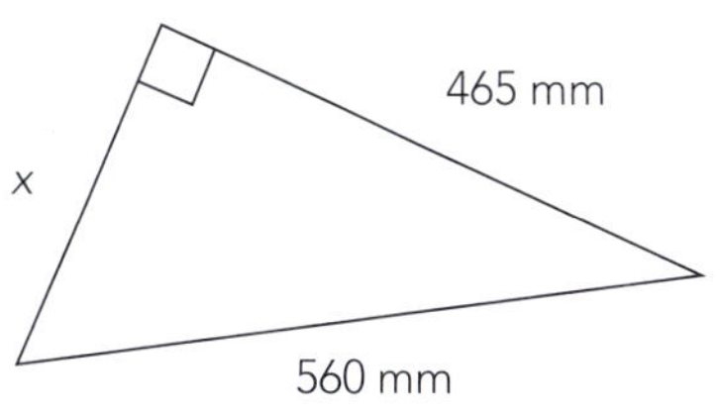

7

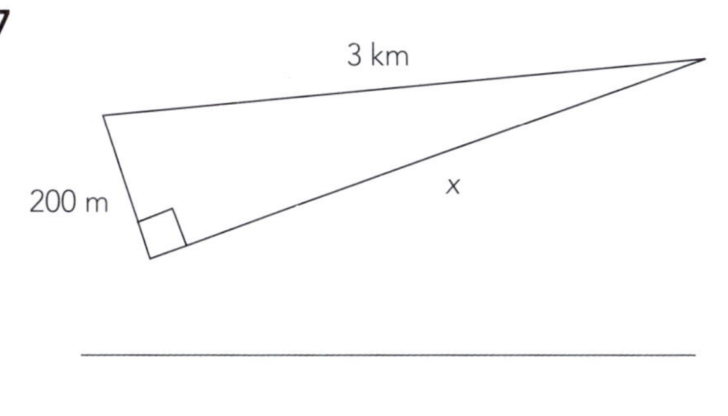

8

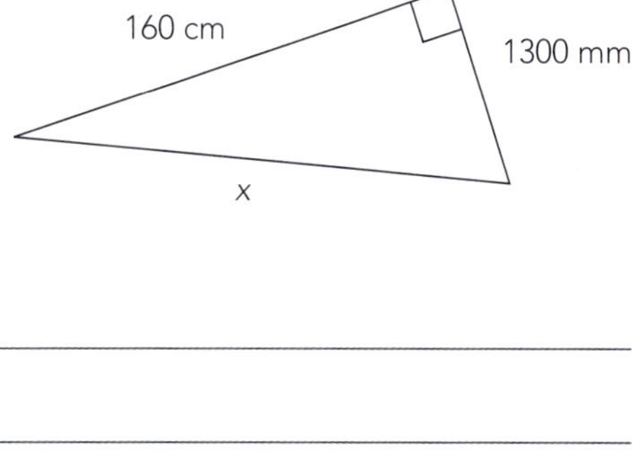

ISBN: 9780170371629

9

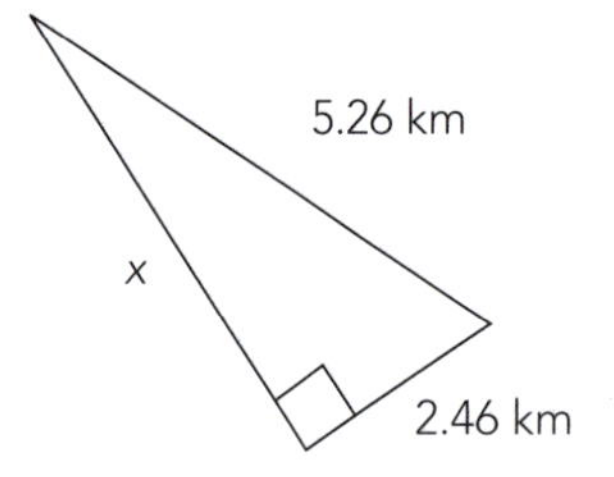

10

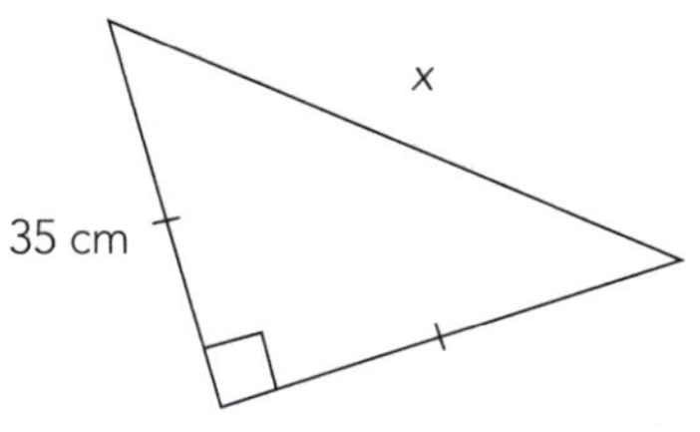

11

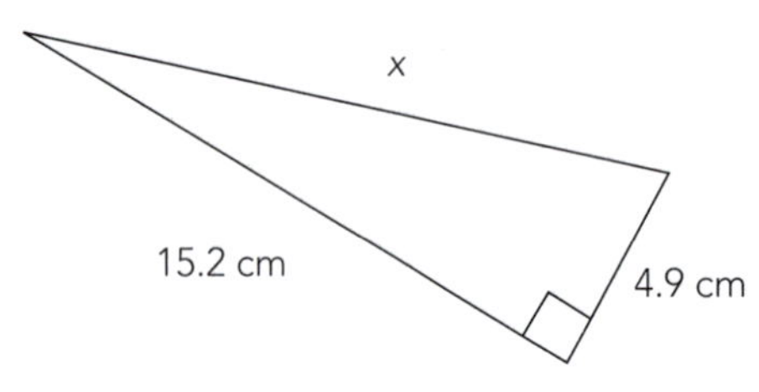

12

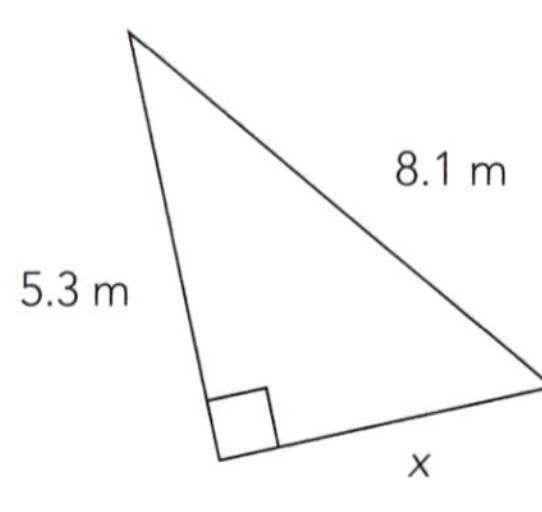

13

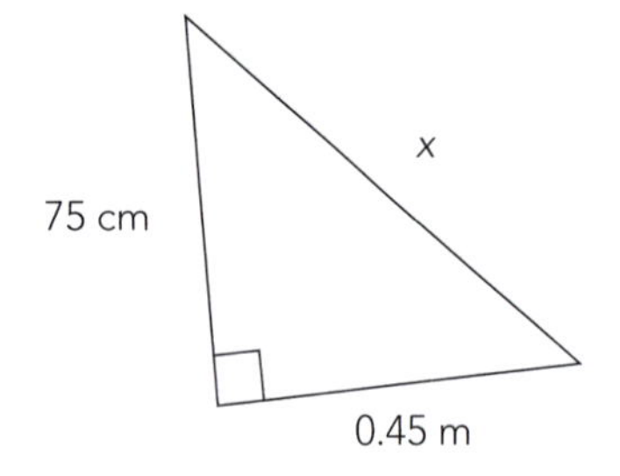

14

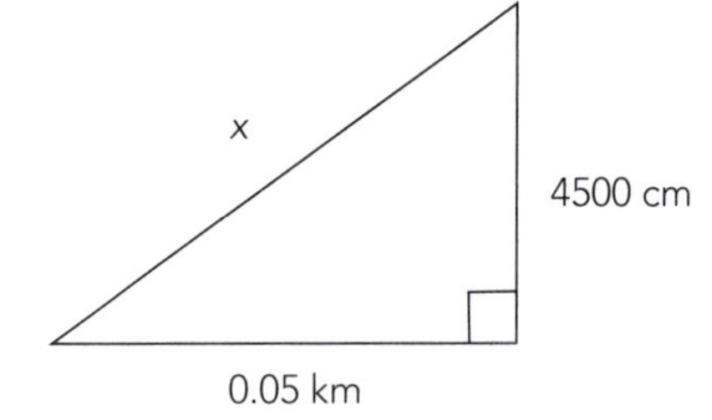

15

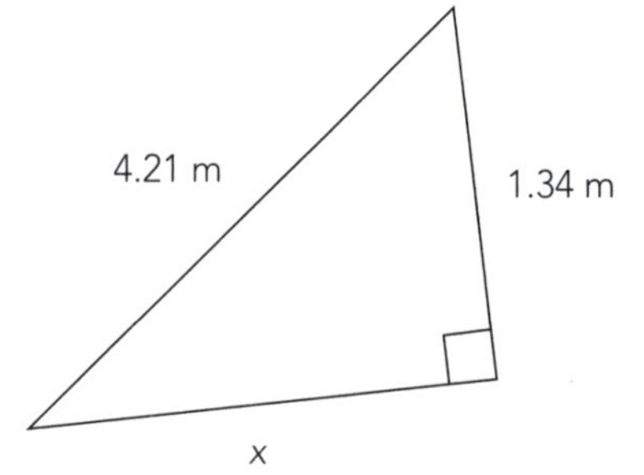

16 The diagram shows an isosceles triangle that has been split into two smaller triangles. Calculate the height (h).

 ISBN: 9780170371629

Applications

Find the missing sides.

1

2 A 4 m long ladder is leaning against a wall. If the base of the ladder is 1.4 m from the wall, calculate how far up the wall (*h*) the ladder reaches.

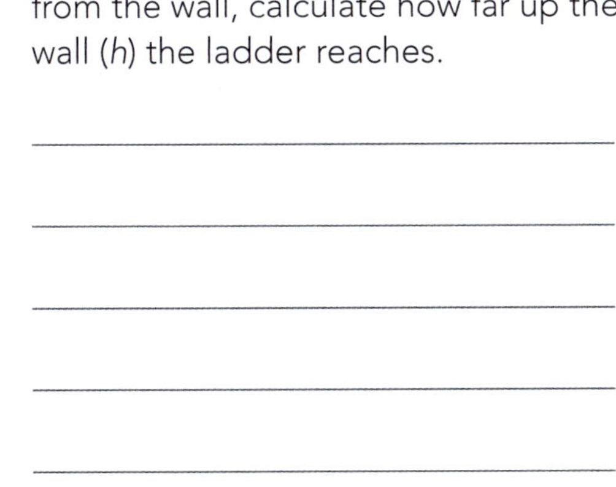

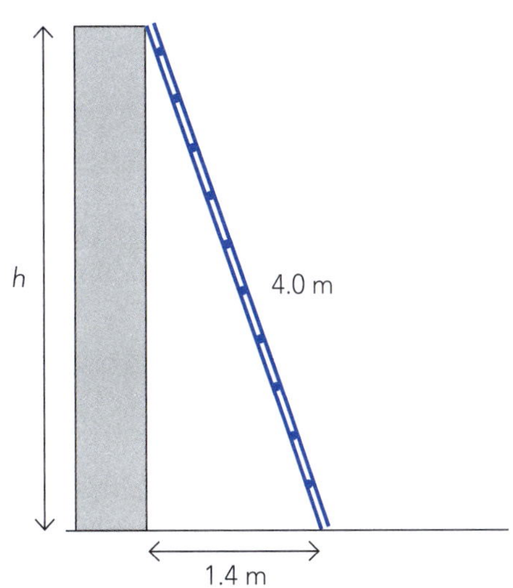

3 ABCD is a trapezium.
Calculate the length of BC.

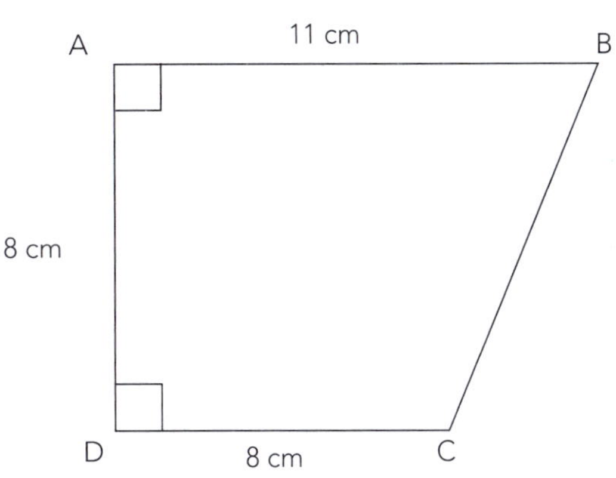

4 ABCD is a trapezium. Calculate the length of AB.

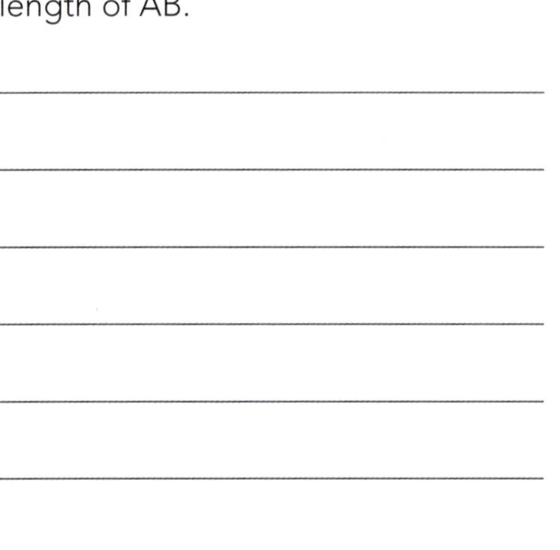

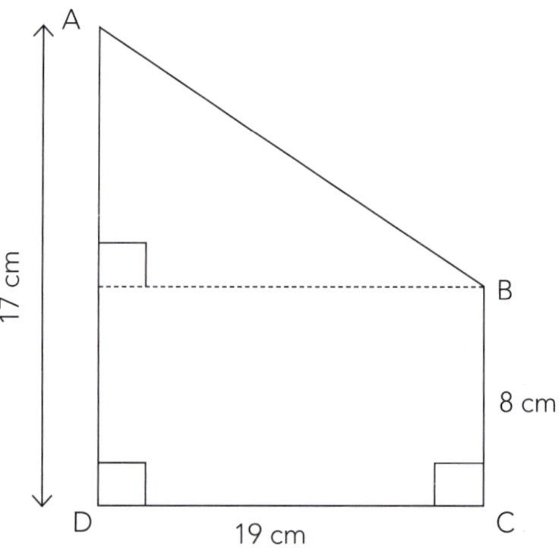

5 Both ABC and ADB are isosceles triangles. Calculate the perimeter of triangle ABC.

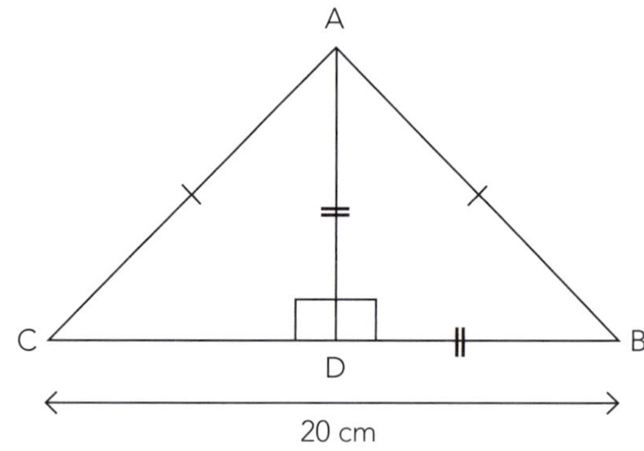

6 Calculate the length of AD.

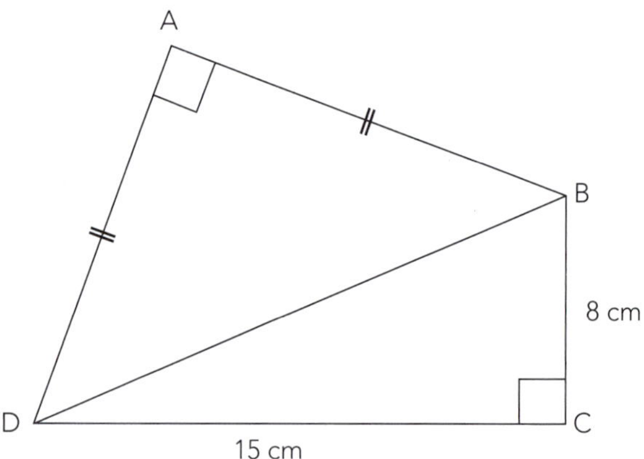

ISBN: 9780170371629

Trigonometry

- Trigonometry can also be used for calculations involving **right-angled triangles**.
- However, unlike calculations using the Theorem of Pythagoras, an **angle** must be involved.

Before you begin a calculation involving trigonometry, **label** the sides of the triangle:

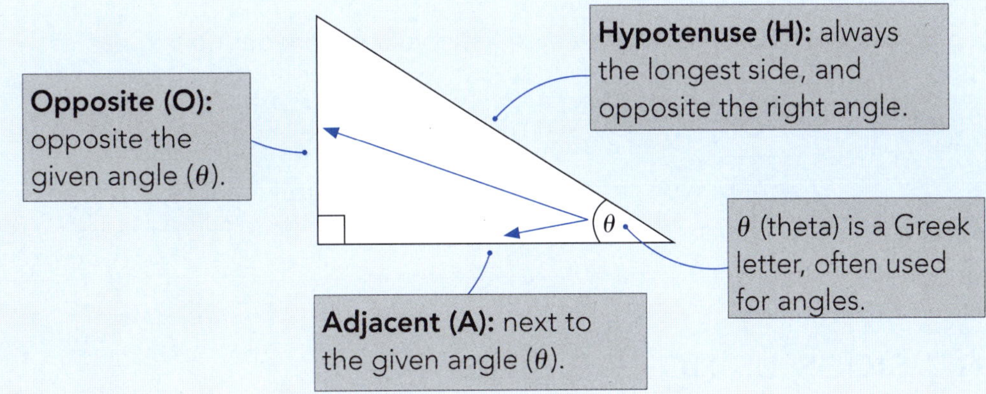

Label the following triangles with H, O and A.

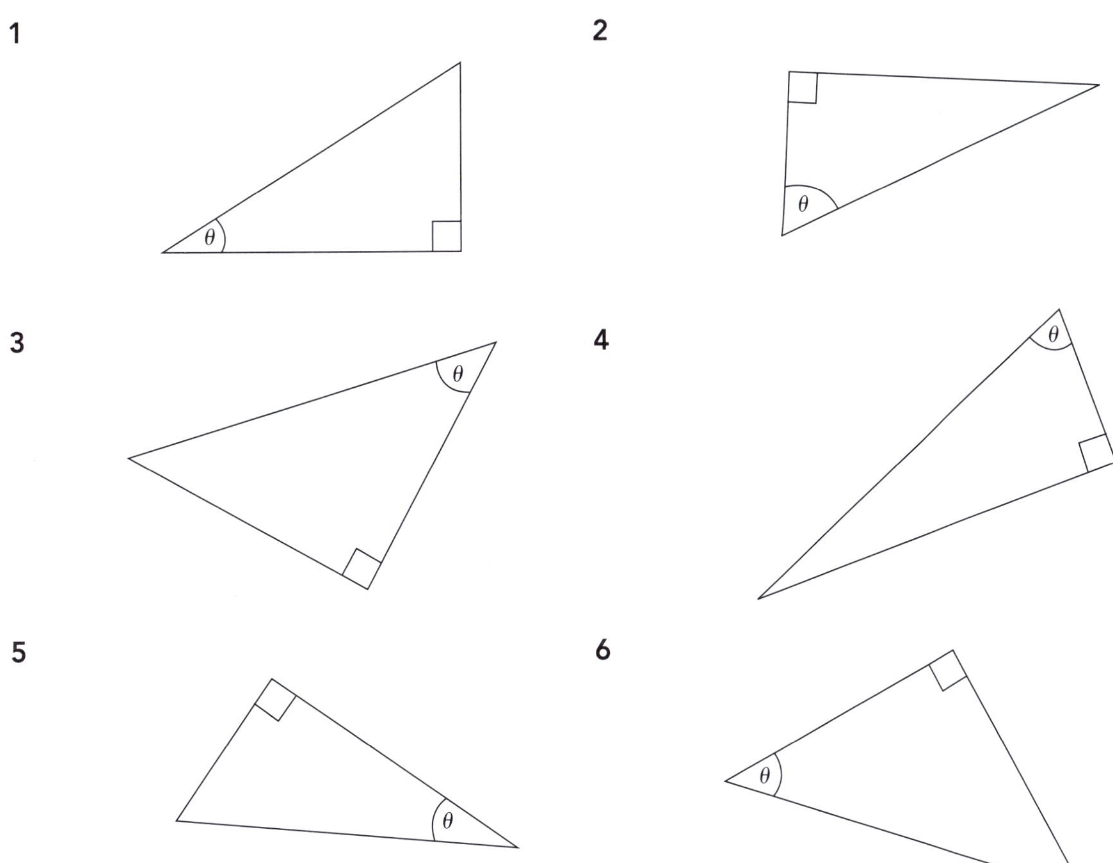

SOH CAH TOA

You will need to know about three trigonometrical functions: **sin** θ (sine)
cos θ (cosine)
tan θ (tangent).

These are the rules in trigonometry:

$$\sin\theta = \frac{O}{H} \qquad \cos\theta = \frac{A}{H} \qquad \tan\theta = \frac{O}{A}$$

These are usually remembered as the 'word' **SOH CAH TOA**.

Organising **SOH CAH TOA** into triangles can help you to work out how to use it:

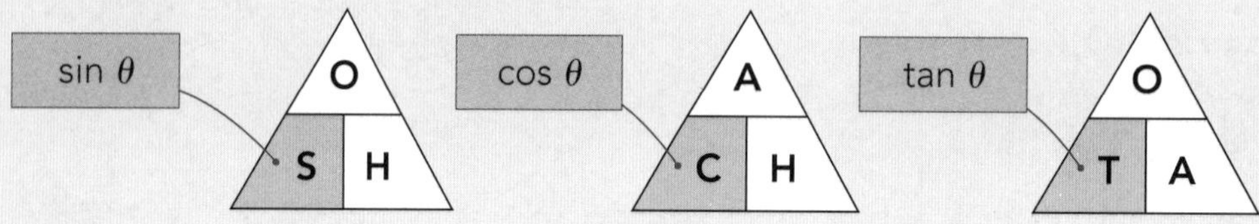

Finding sides using the sine

Example one: Calculate the length marked *x*.

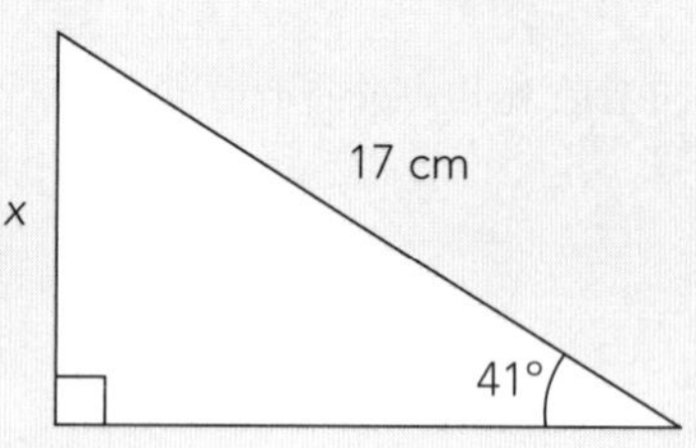

Step 1: **Label** the sides that are involved in the problem with 'A', 'O' and 'H'.

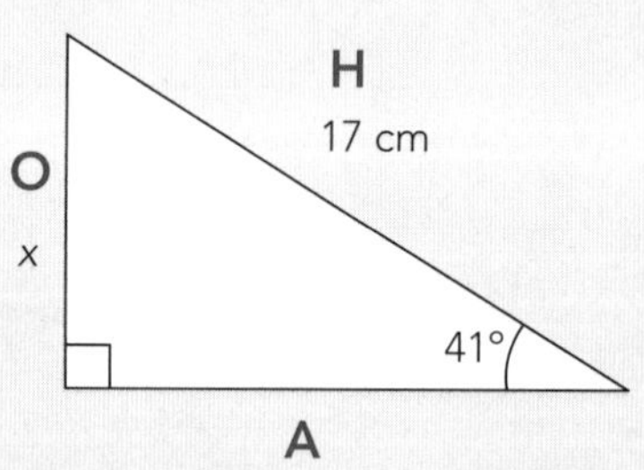

Step 2: Decide **which relationship** involves these sides. In this case we are concerned only with O and H so we must use:

$$\sin\theta = \frac{O}{H}$$

Step 3: **Substitute** the values from the triangle.

$$\sin 41° = \frac{x}{17}$$

$$x = 17\sin 41°$$
$$= 11.15 \text{ cm}$$

Step 4: **Think about your answer — does it seem reasonable?**
In this case, the hypotenuse is 17 cm and *x* must be shorter, so an answer of 11.15 cm is reasonable.

ISBN: 9780170371629

This is the same process, but the algebra is a little different.

Example two: Calculate the length marked x.

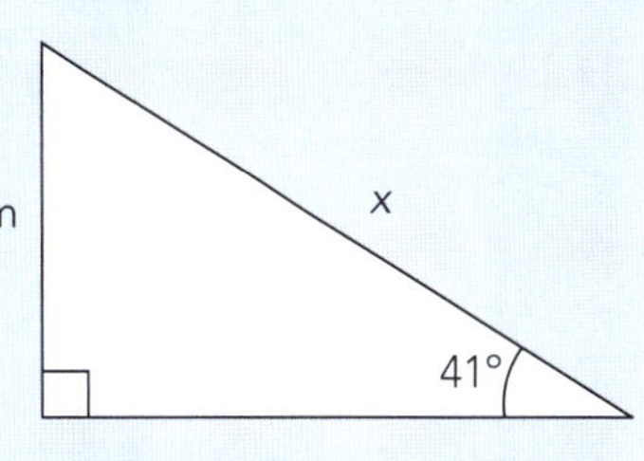

Step 1: **Label** the sides that are involved in the problem with 'A', 'O' and 'H'.

Step 2: Decide **which relationship** involves these sides. In this case we are concerned only with O and H so we must use:

$$\sin\theta = \frac{O}{H}$$

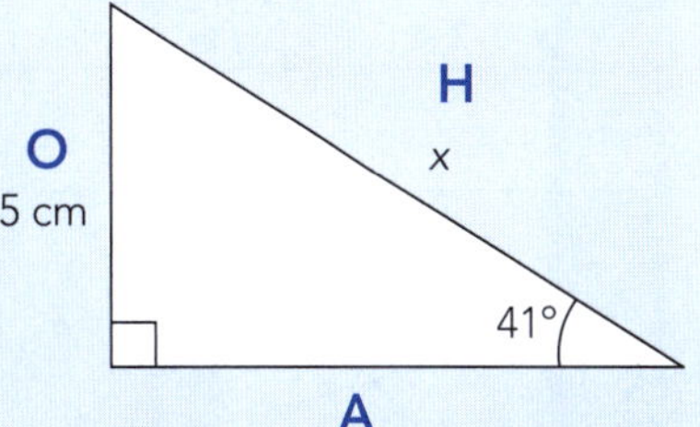

Step 3: **Substitute** the values from the triangle.

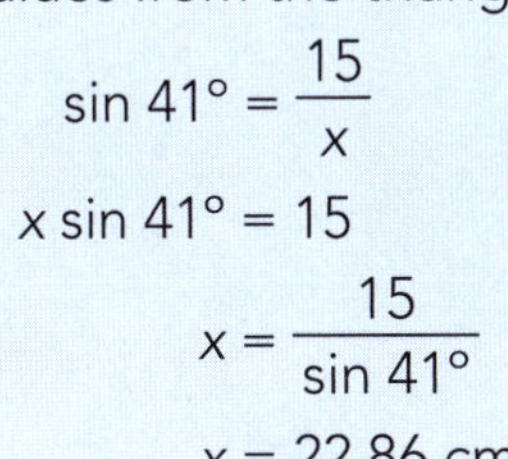

$$\sin 41° = \frac{15}{x}$$

$$x \sin 41° = 15$$

$$x = \frac{15}{\sin 41°}$$

$$x = 22.86 \text{ cm}$$

Step 4: **Think about your answer — does it seem reasonable?** In this case, the opposite side is 15 cm and x must be longer, so an answer of 22.86 cm is reasonable.

Find the missing sides of these triangles.

1

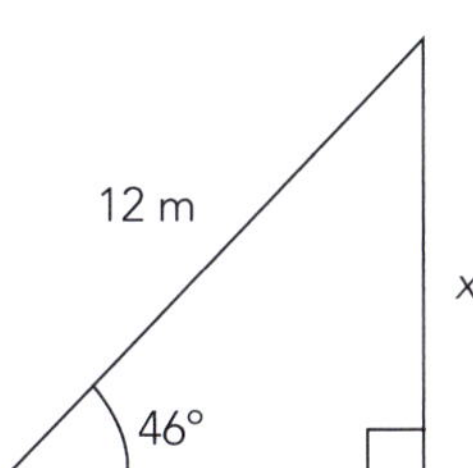

2

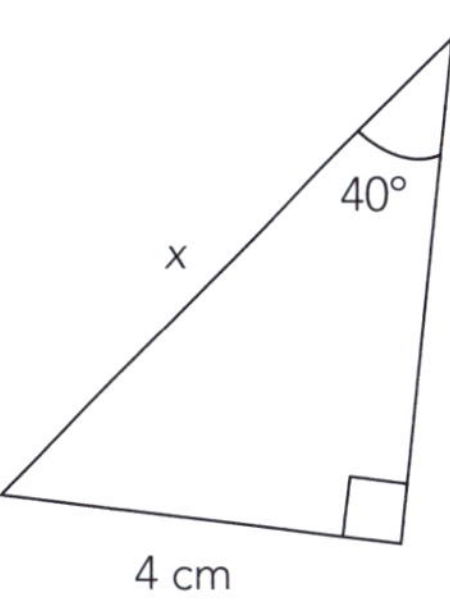

3

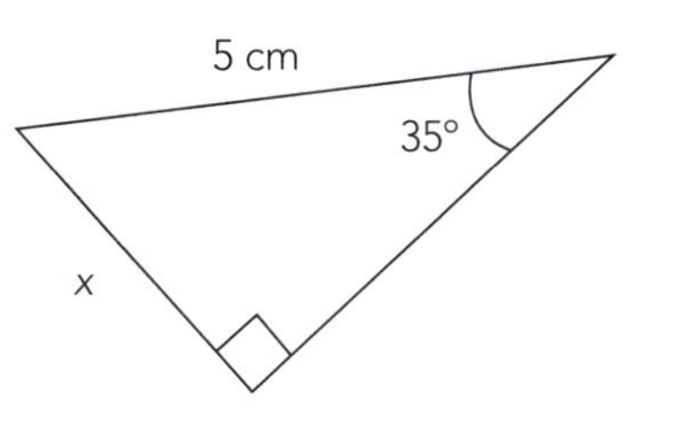

4

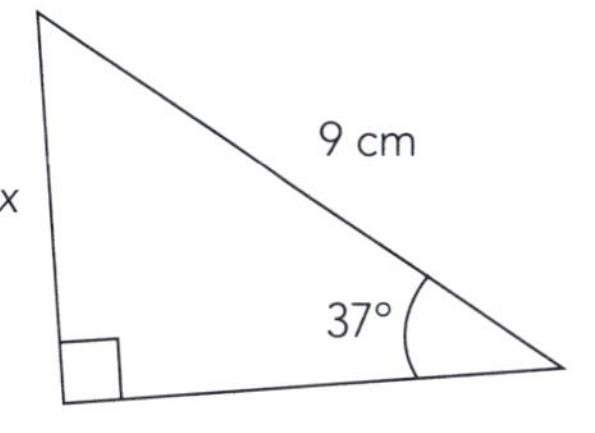

ISBN: 9780170371629

5

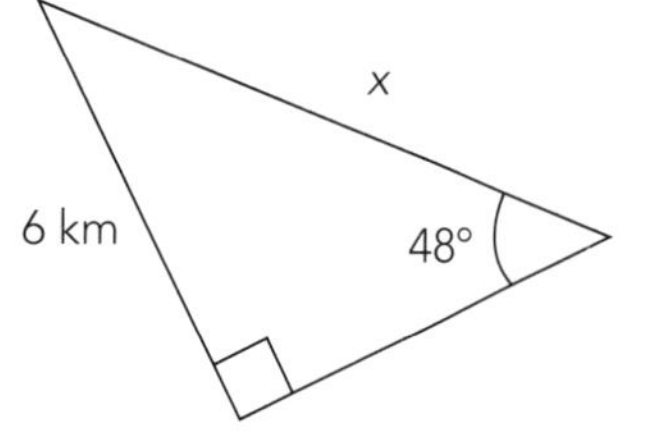

6

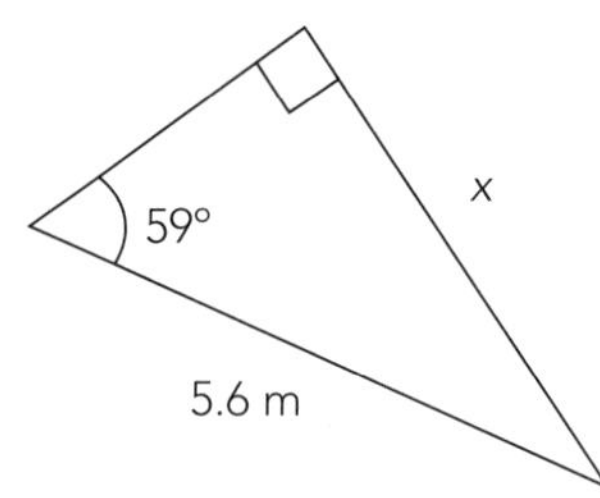

7

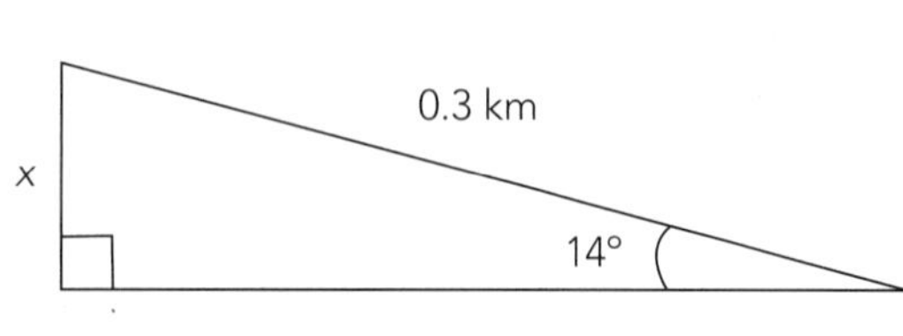

8

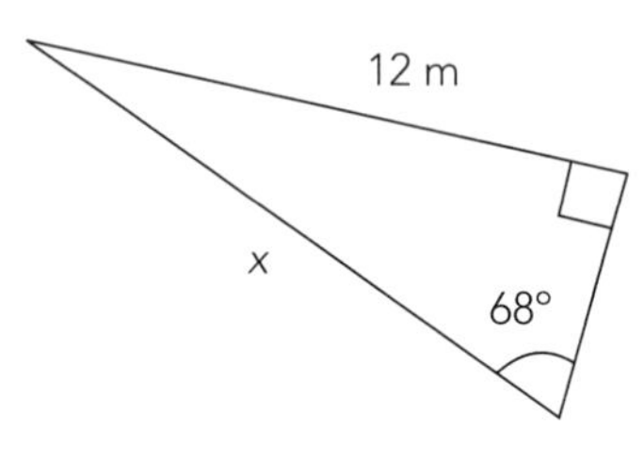

9

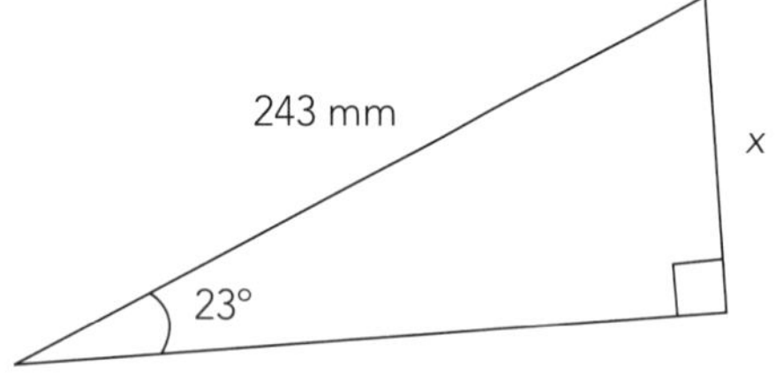

10

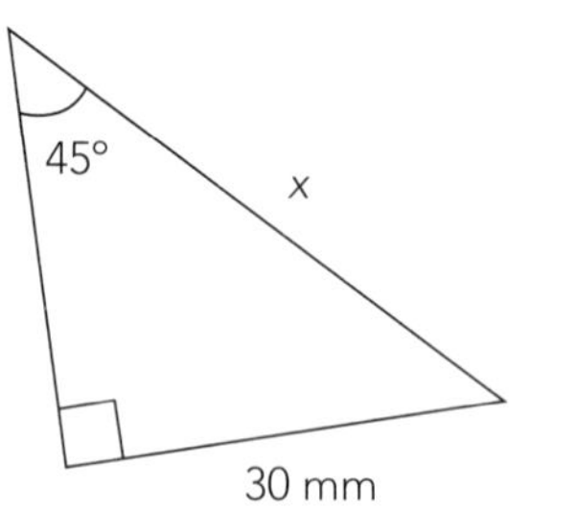

11

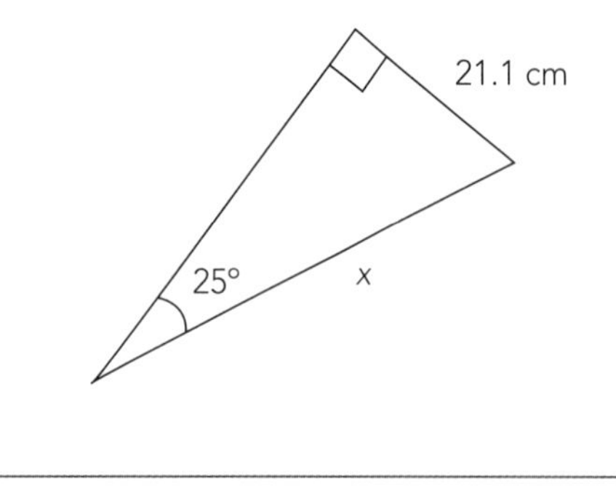

12

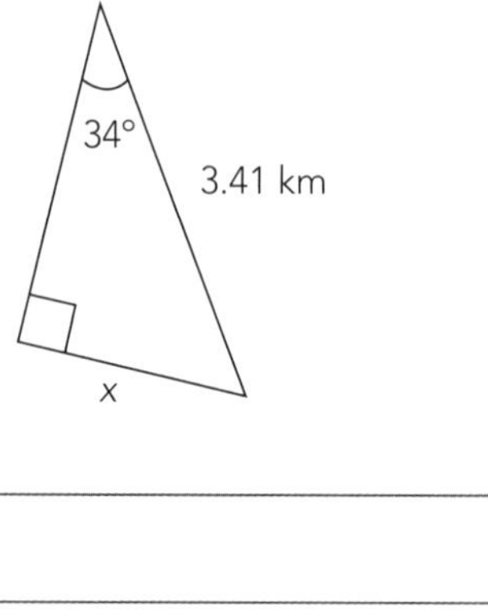

ISBN: 9780170371629

13

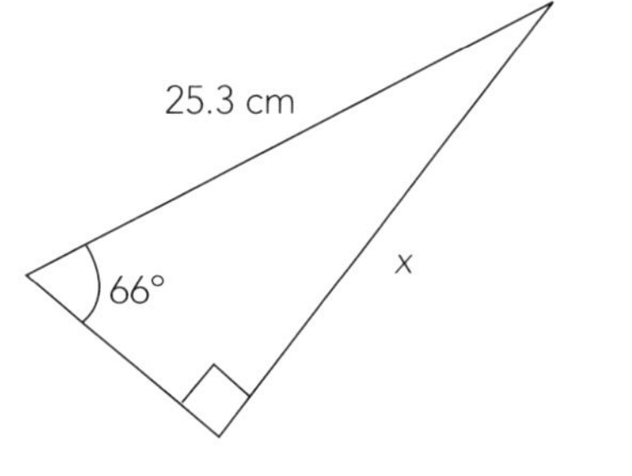

14

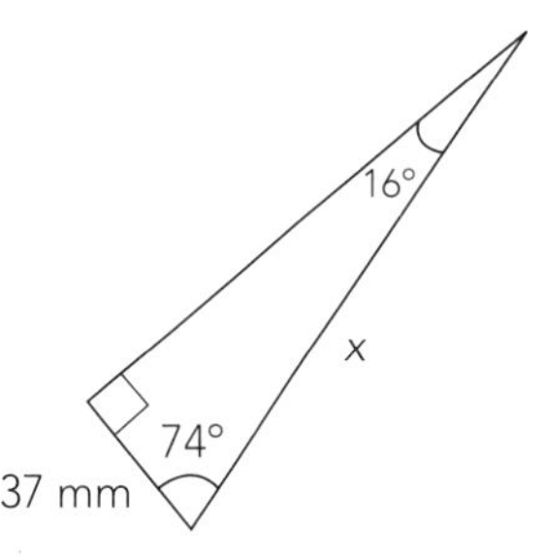

15

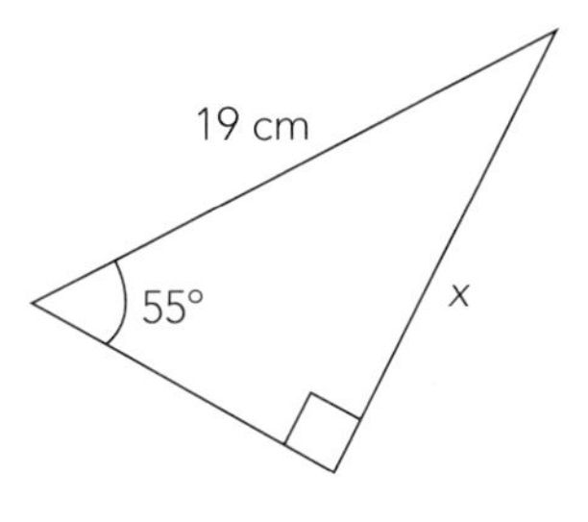

16

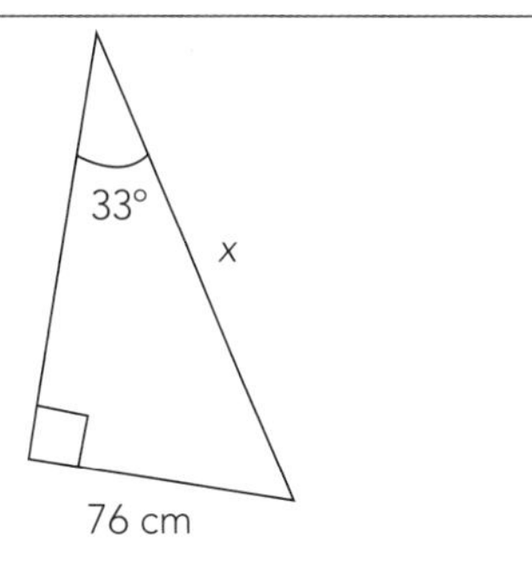

17

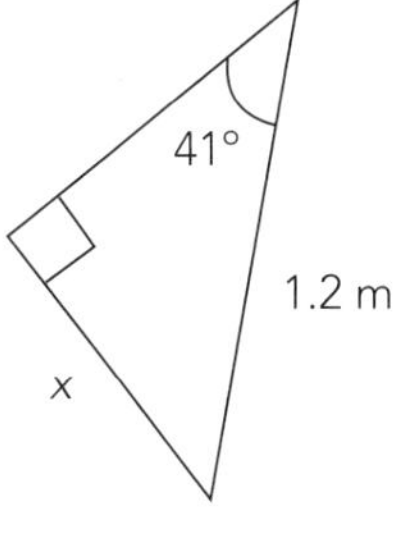

18

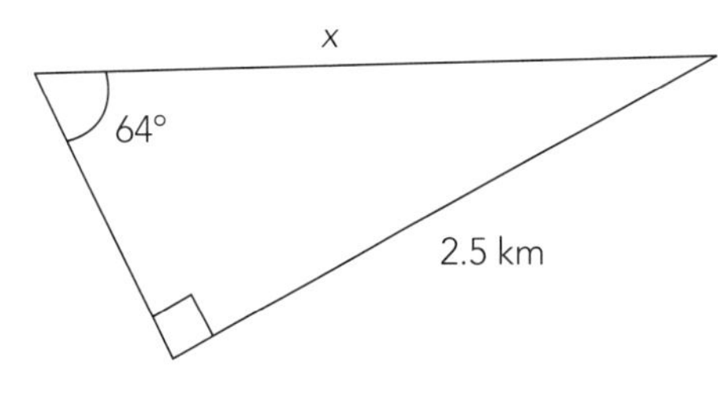

19

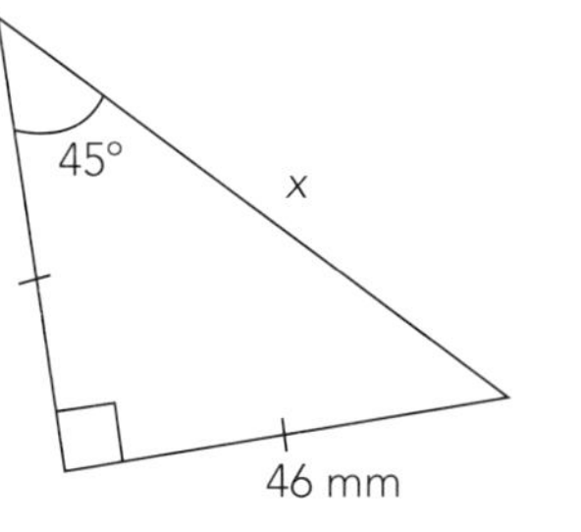

20

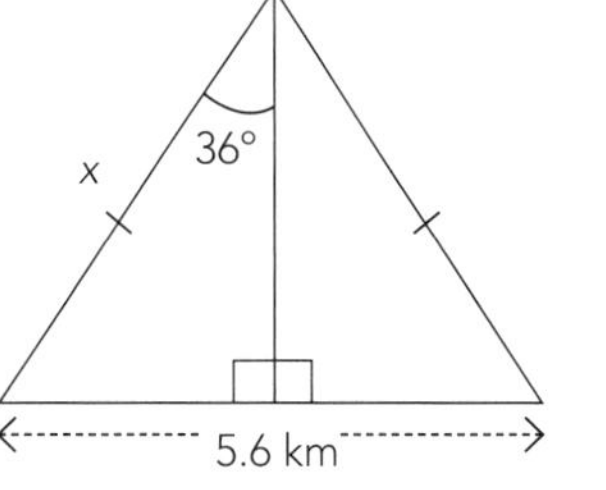

ISBN: 9780170371629

Finding sides using the cosine

Example one: Calculate the length marked x.

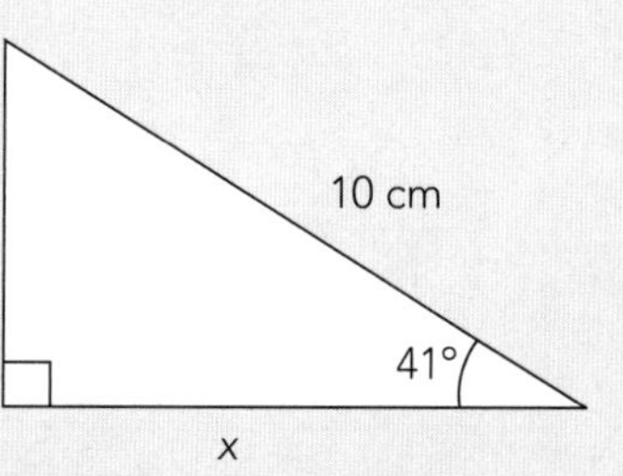

Step 1: **Label** the sides that are involved in the problem with 'A', 'O' and 'H'.

Step 2: Decide **which relationship** involves these sides. In this case we are concerned only with A and H so we must use:

$$\cos\theta = \frac{A}{H}$$

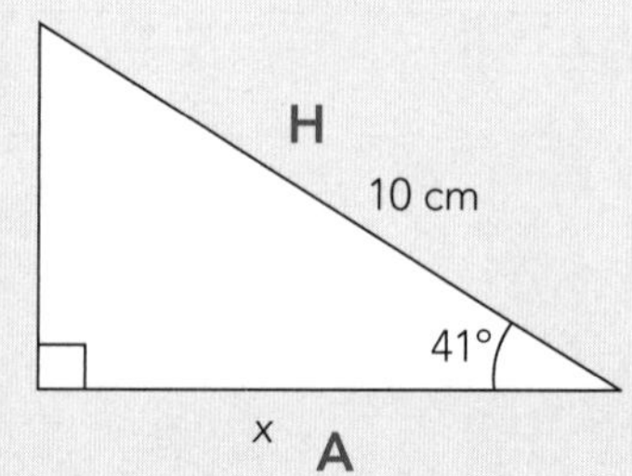

Step 3: **Substitute** the values from the triangle.

$$\cos 41° = \frac{x}{10}$$

$$x = 10\cos 41°$$

$$= 7.547 \text{ cm}$$

> Check that your calculator is set to **degrees**.

Step 4: **Think about your answer — does it seem reasonable?**
In this case, the hypotenuse is 10 cm and x must be shorter, so an answer of 7.547 cm is reasonable.

This is the same process, but the algebra is a little different.

Example two: Calculate the length marked x.

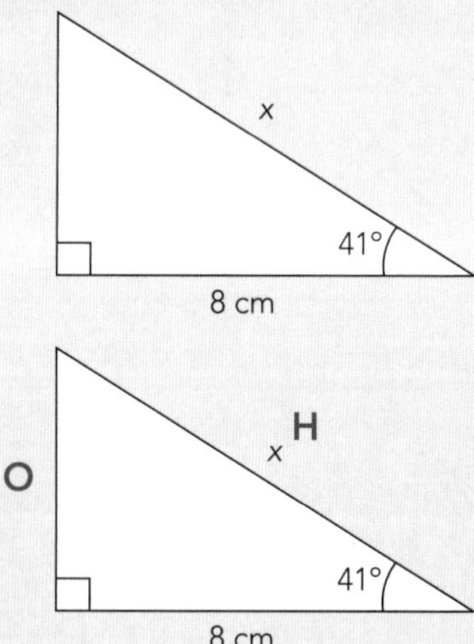

Step 1: **Label** the sides that are involved in the problem with 'A', 'O' and 'H'.

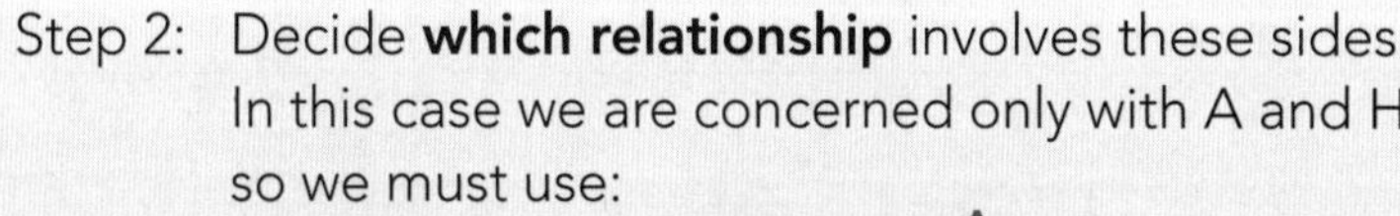

Step 2: Decide **which relationship** involves these sides. In this case we are concerned only with A and H so we must use:

$$\cos\theta = \frac{A}{H}$$

Step 3: **Substitute** the values from the triangle.

$$\cos 41° = \frac{8}{x}$$

$$x\cos 41° = 8$$

$$x = \frac{8}{\cos 41°}$$

$$x = 10.60 \text{ cm}$$

Step 4: **Think about your answer — does it seem reasonable?**
In this case, the adjacent side is 8 cm and x must be longer, so an answer of 10.60 cm is reasonable.

ISBN: 9780170371629

Calculate the unknown sides of the following triangles.

1

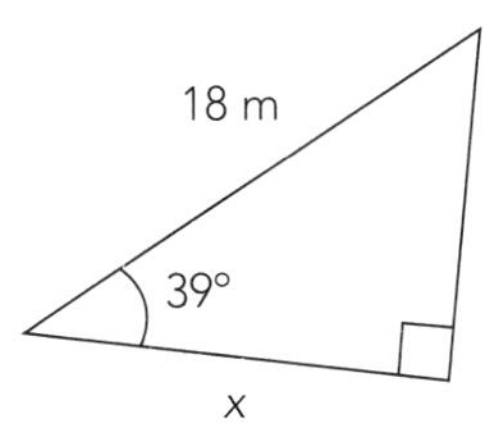

2

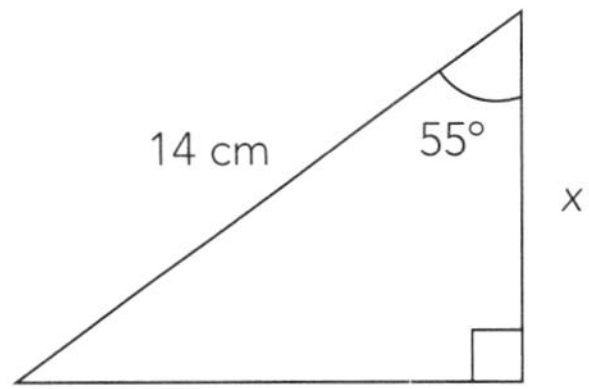

3

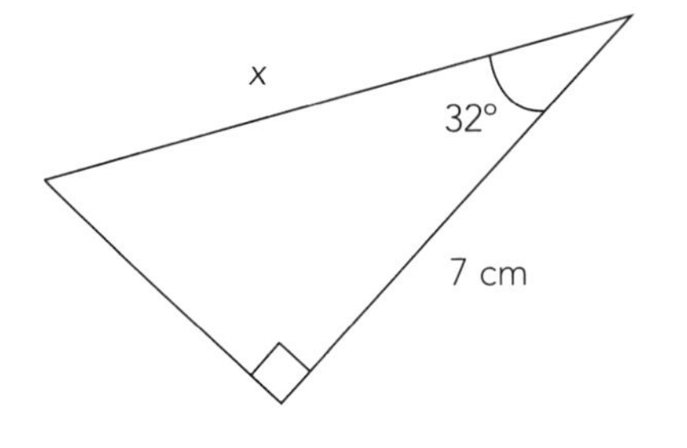

4

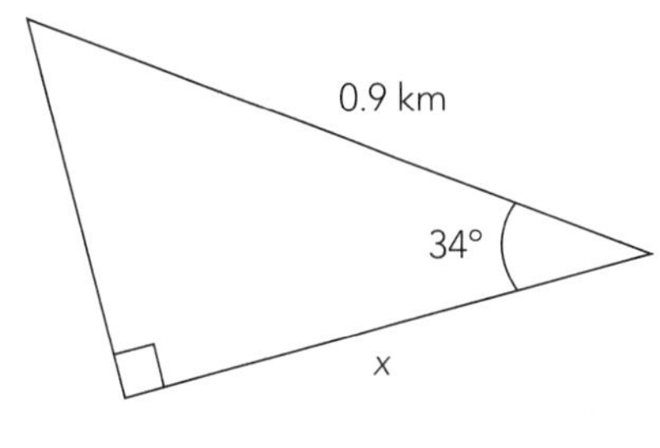

5

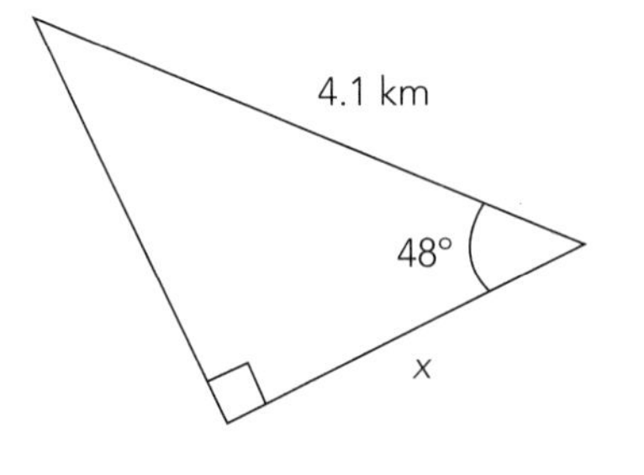

6

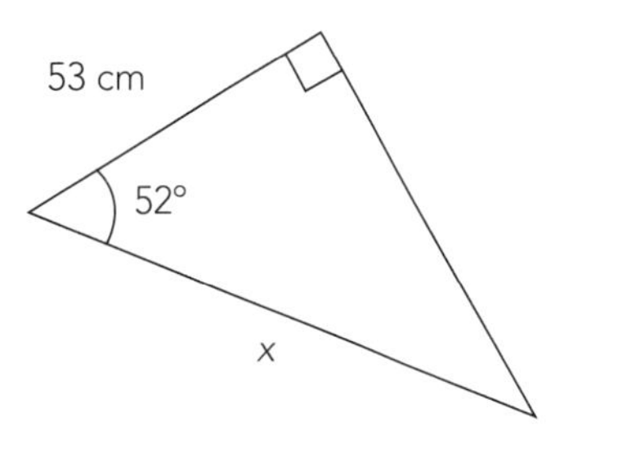

7

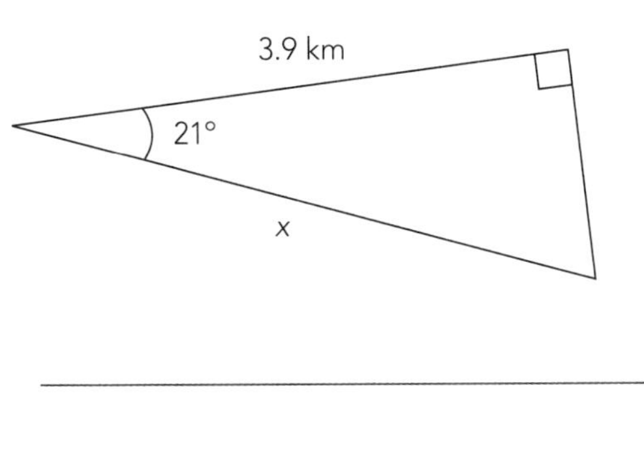

8

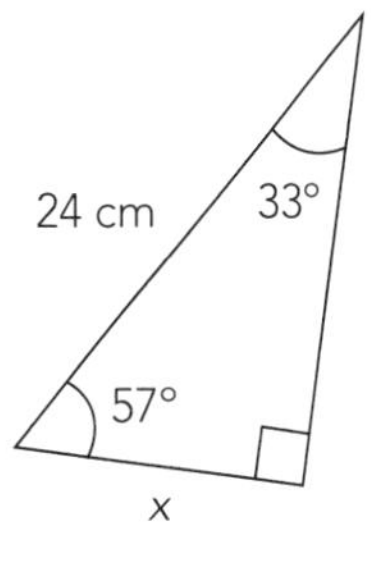

ISBN: 9780170371629

9

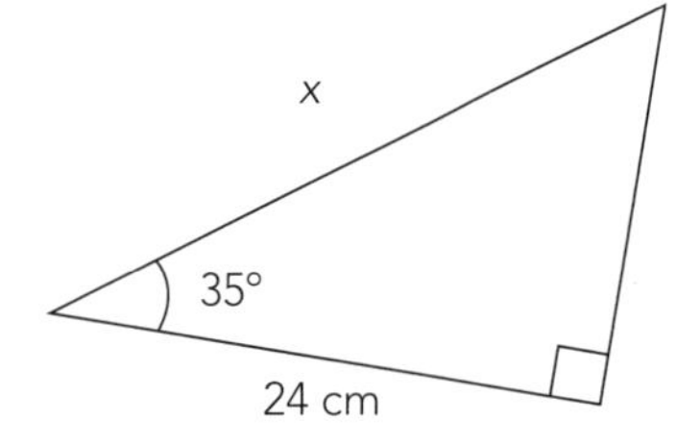

10

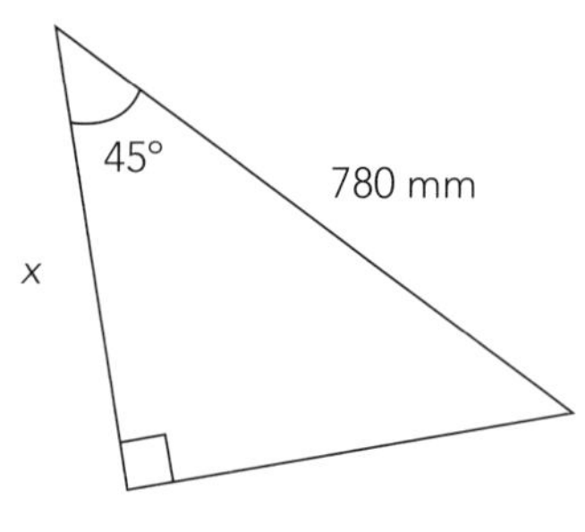

11

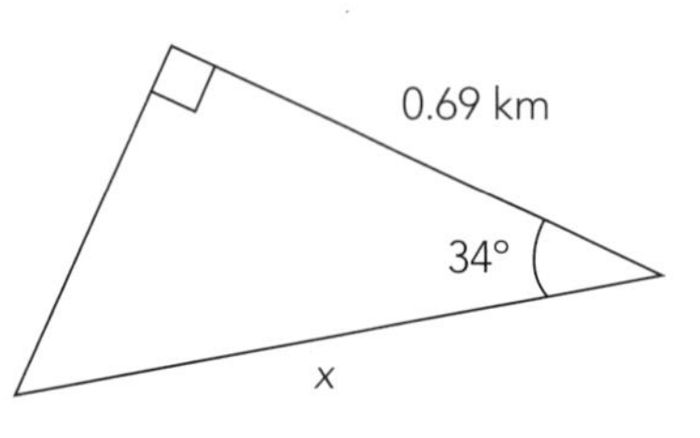

12

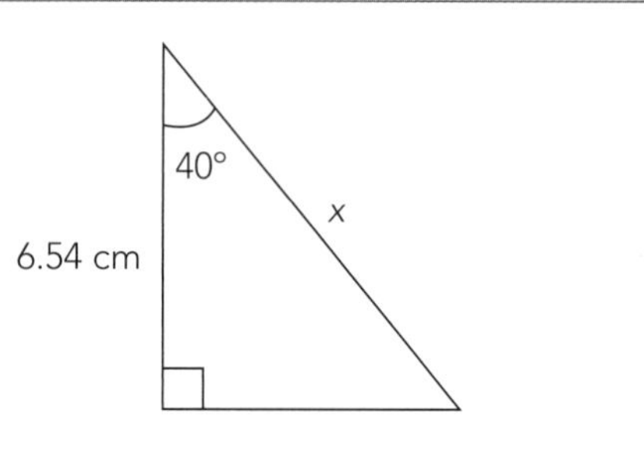

13

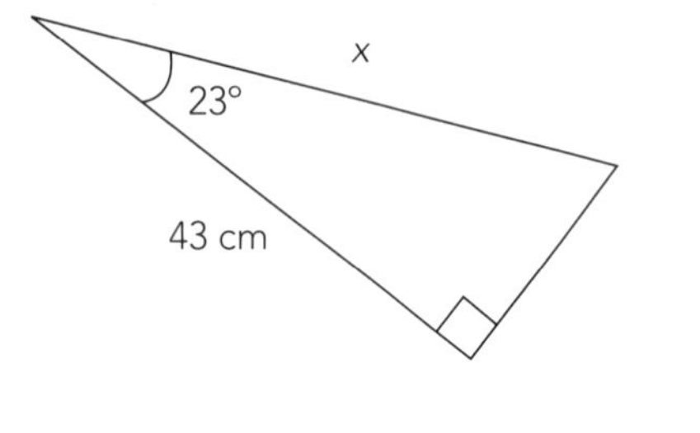

14

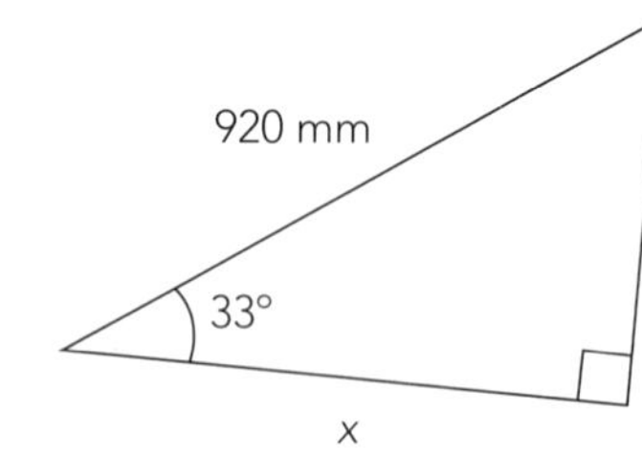

15

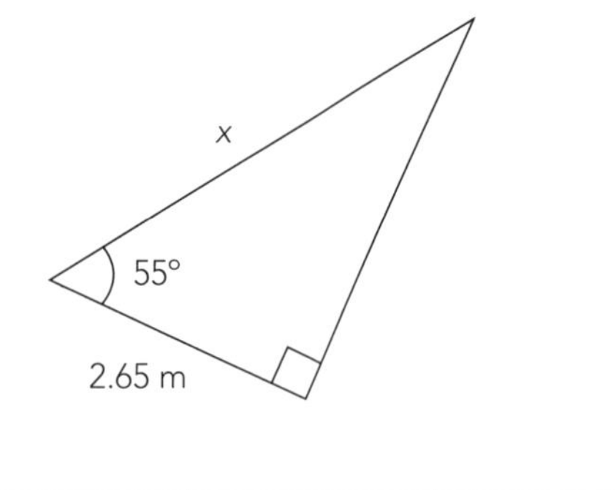

16

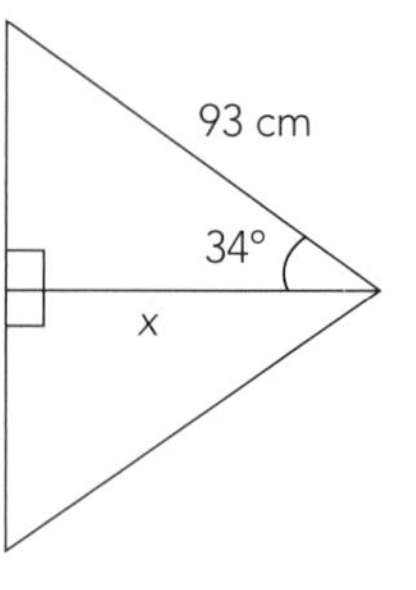

 ISBN: 9780170371629

Finding sides using the tangent

This involves using either of the algebraic methods used earlier.

Example one: Calculate the length marked x.

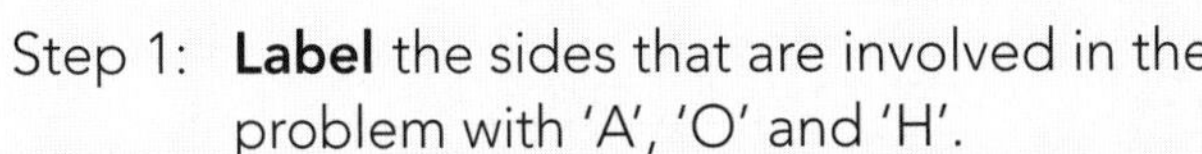

Step 1: **Label** the sides that are involved in the problem with 'A', 'O' and 'H'.

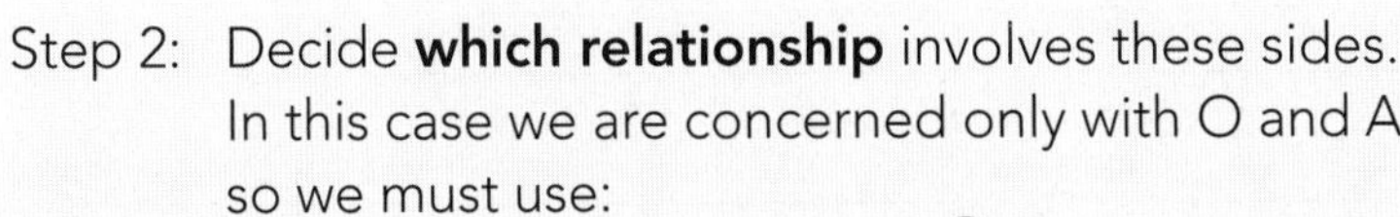

Step 2: Decide **which relationship** involves these sides. In this case we are concerned only with O and A so we must use:

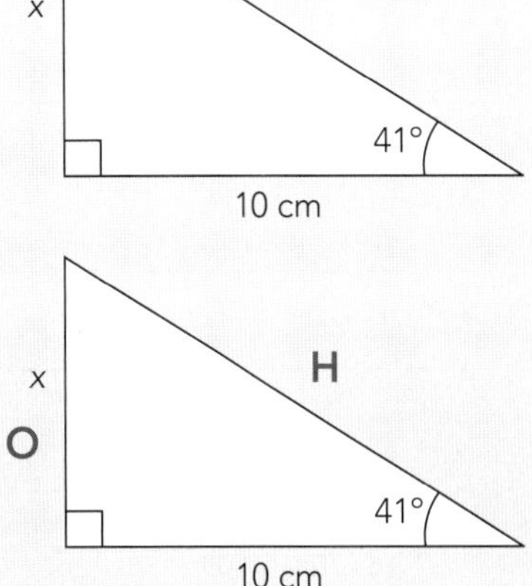

$$\tan \theta = \frac{\mathbf{O}}{\mathbf{A}}$$

Step 3: **Substitute** the values from the triangle.

$$\tan 41° = \frac{x}{10}$$

$$x = 10 \tan 41°$$

$$x = 8.693 \text{ cm}$$

Step 4: **Think about your answer — does it seem reasonable?** In this case, x must be less than 10 cm because x is opposite the smallest angle (41°), so an answer of 8.693 cm is reasonable.

This is the same process, but the algebra is a little different.

Example two: Calculate the length marked x.

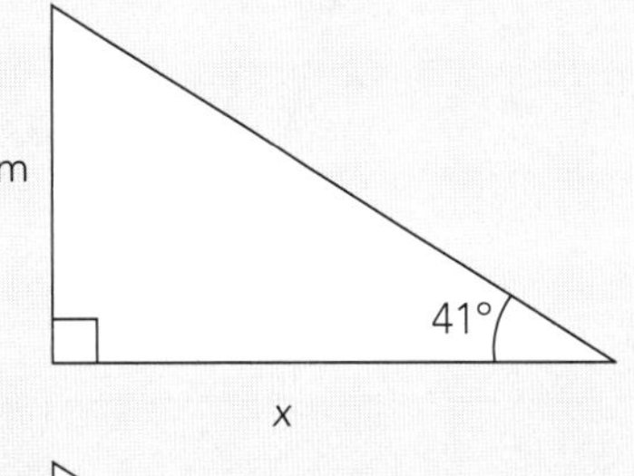

Step 1: **Label** the sides that are involved in the problem with 'A', 'O' and 'H'.

Step 2: Decide **which relationship** involves these sides. In this case we are concerned only with O and A so we must use:

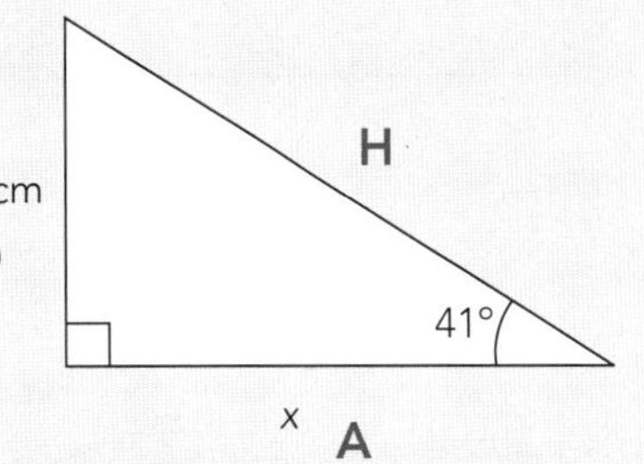

$$\tan \theta = \frac{\mathbf{O}}{\mathbf{A}}$$

Step 3: **Substitute** the values from the triangle.

$$\tan 41° = \frac{17}{x}$$

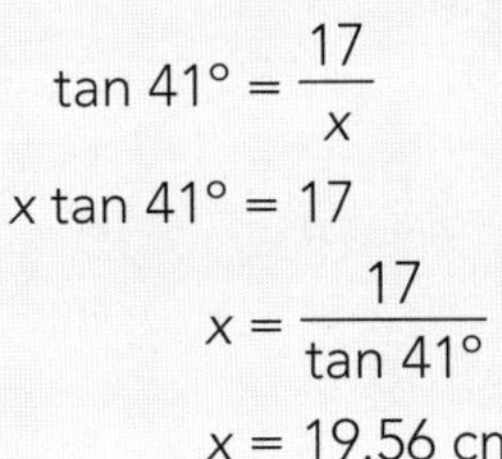

$$x \tan 41° = 17$$

$$x = \frac{17}{\tan 41°}$$

$$x = 19.56 \text{ cm}$$

Step 4: **Think about your answer — does it seem reasonable?** In this case, x must be more than 17 cm because the 17 cm is opposite the smallest angle (41°), so an answer of 19.56 cm is reasonable.

ISBN: 9780170371629

Calculate the unknown sides of the following triangles.

1

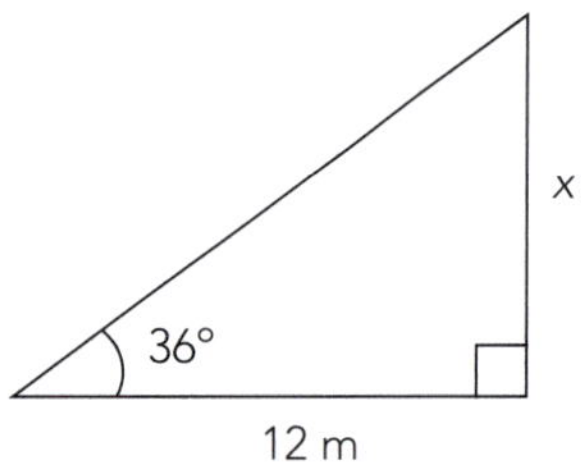

2

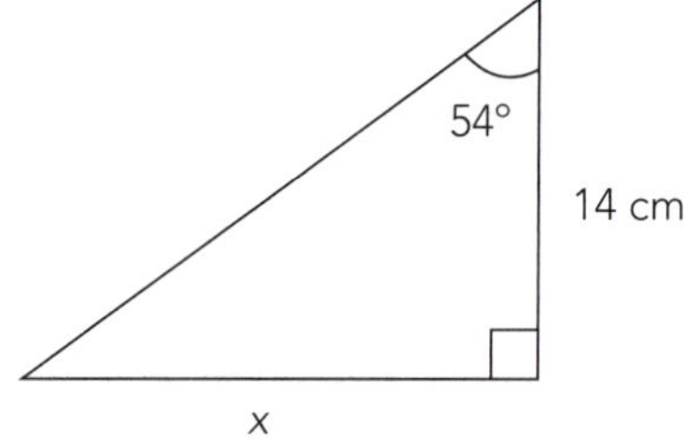

3

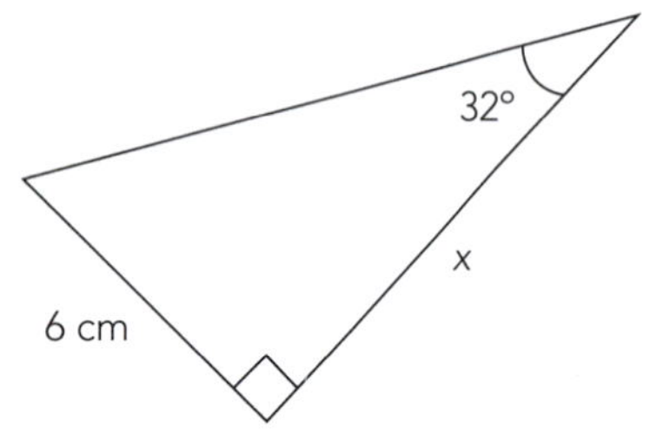

4

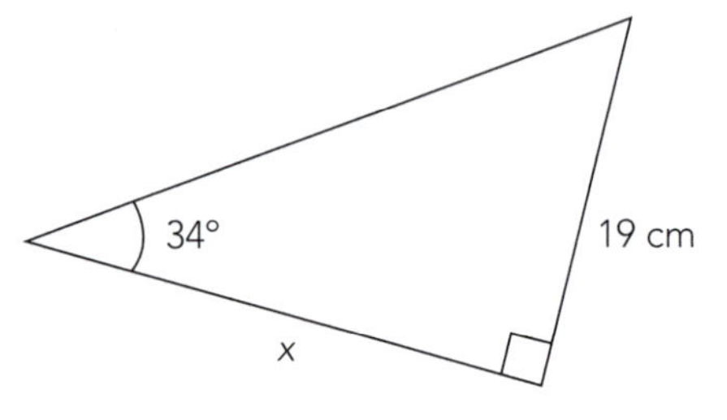

5

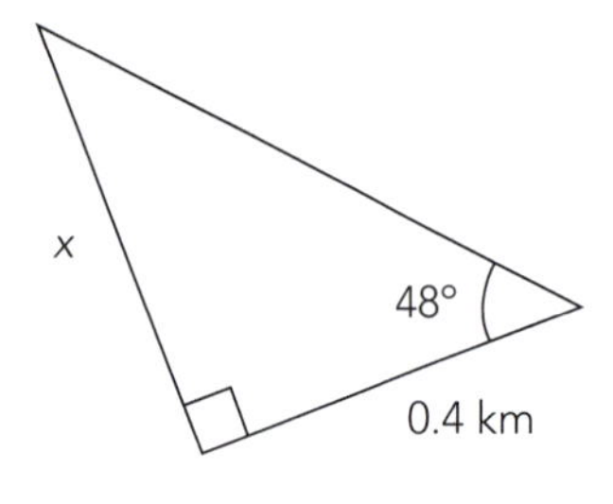

6

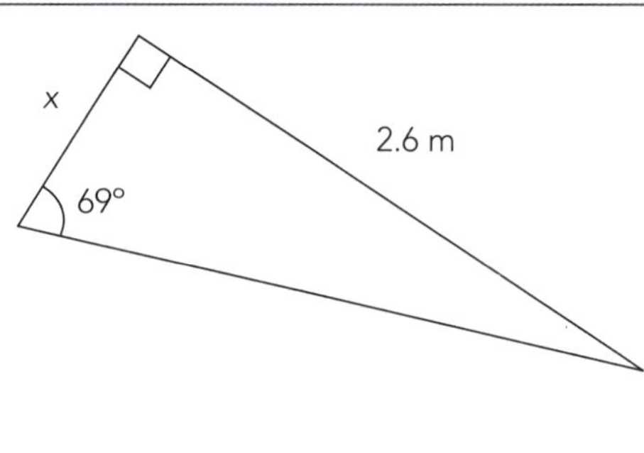

7

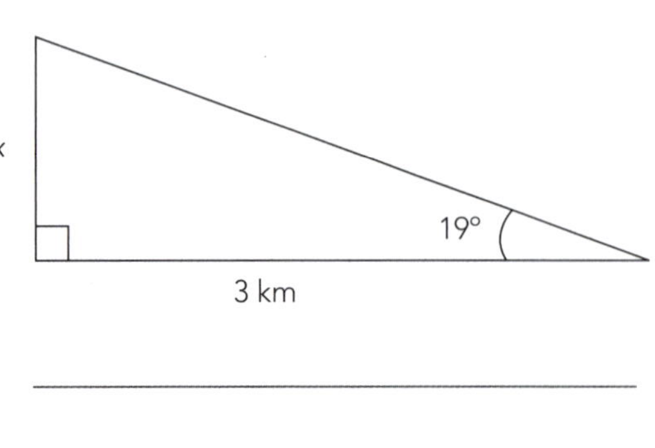

8

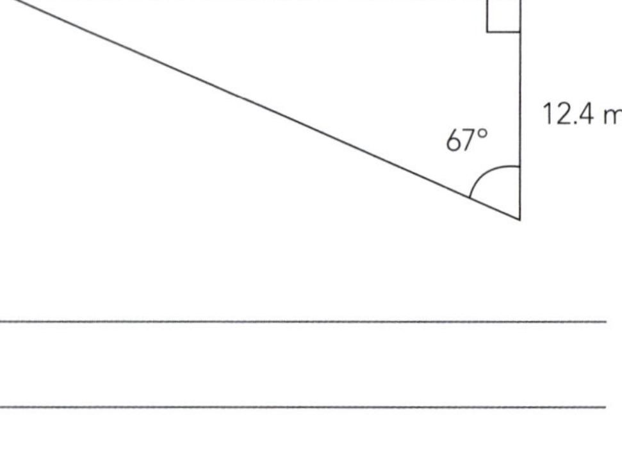

ISBN: 9780170371629

9

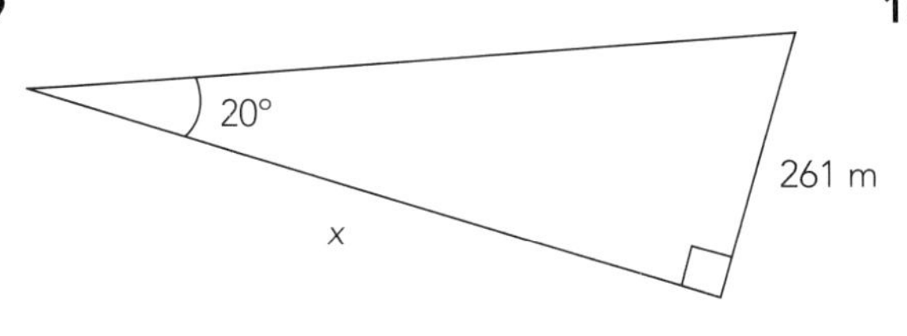

10

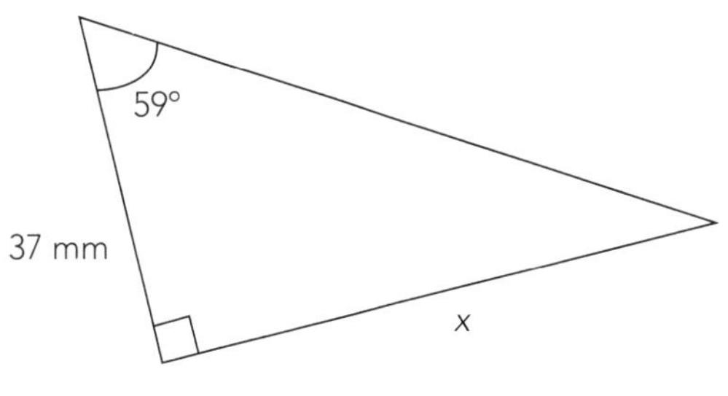

11

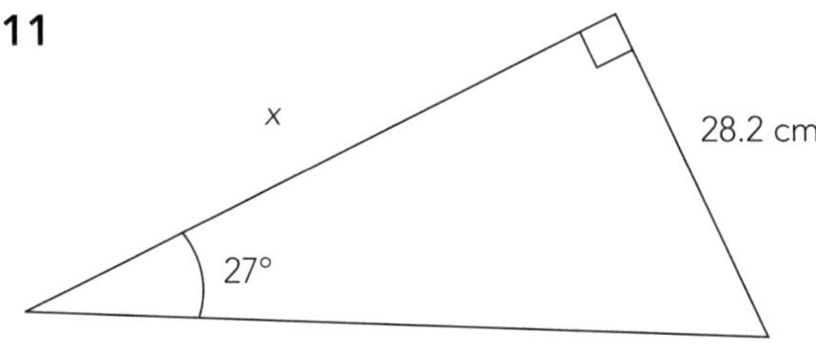

12

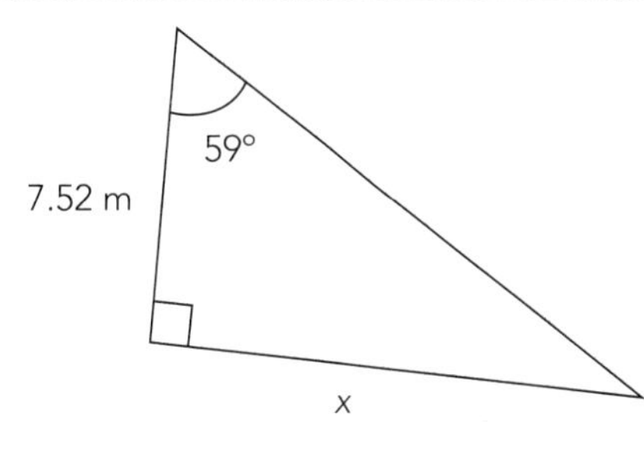

13

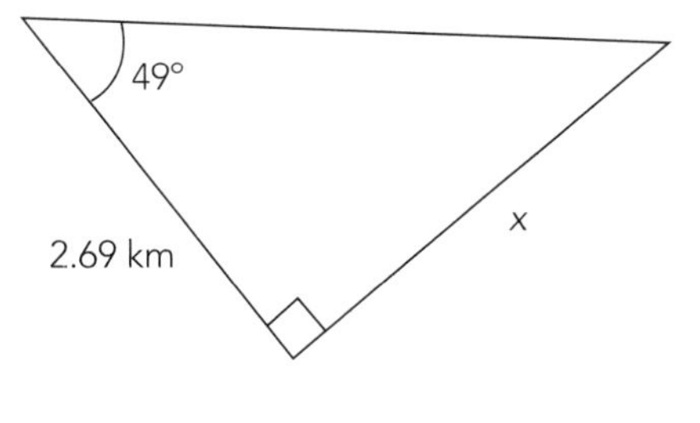

14

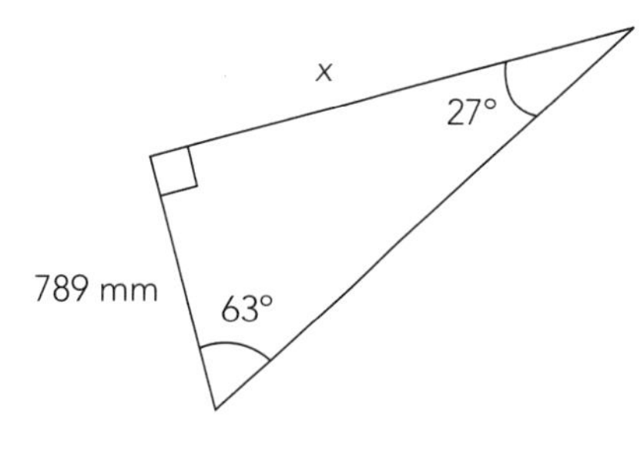

15

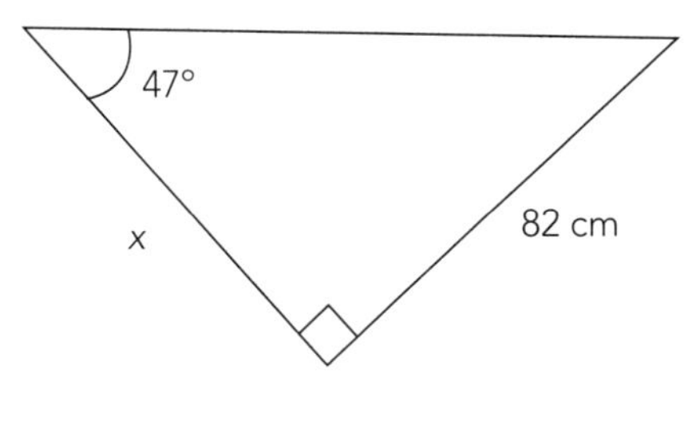

16

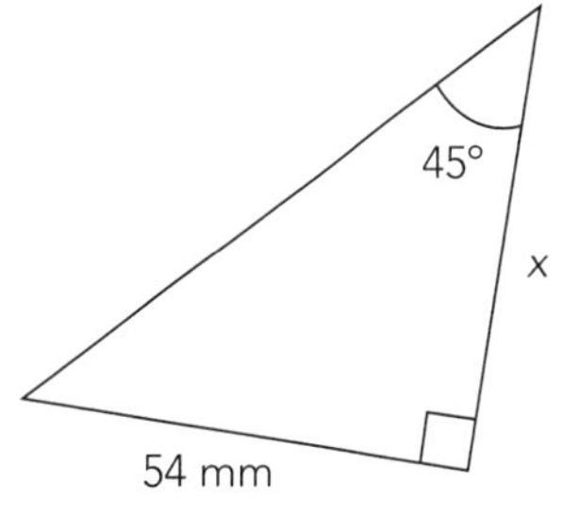

ISBN: 9780170371629

Finding sides — mixing it up

1

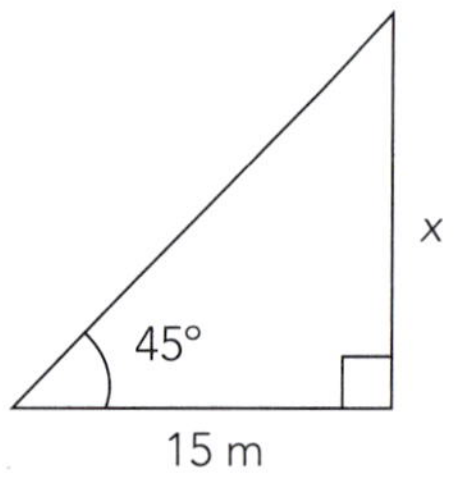

2

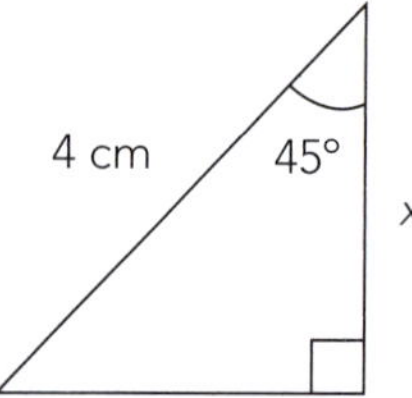

3

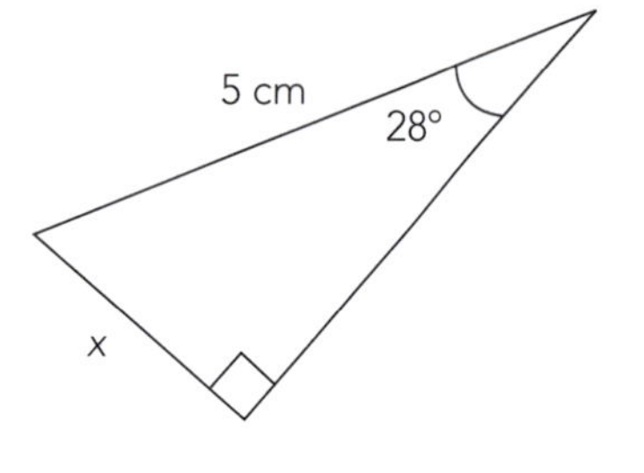

4

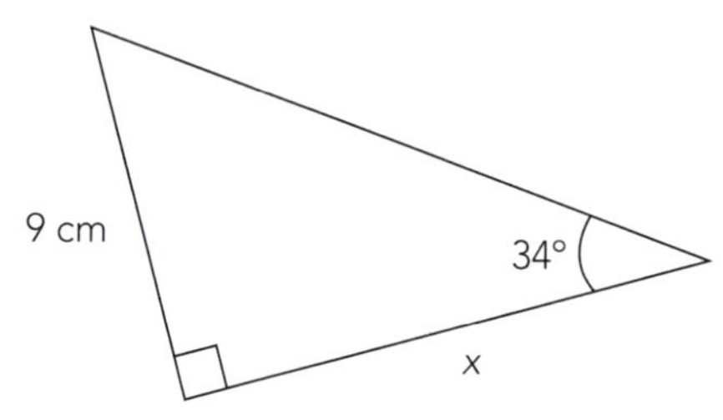

5

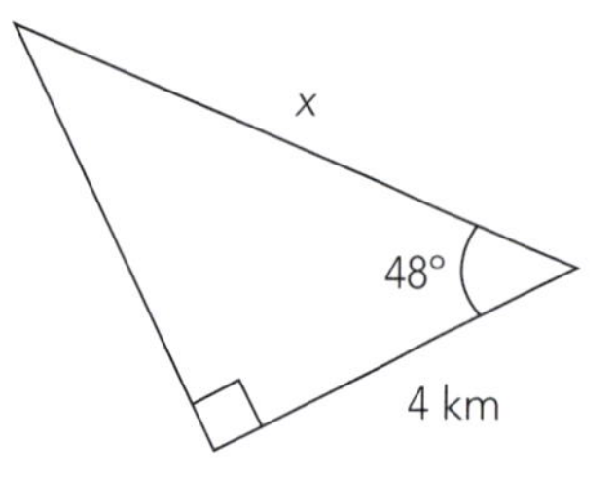

6

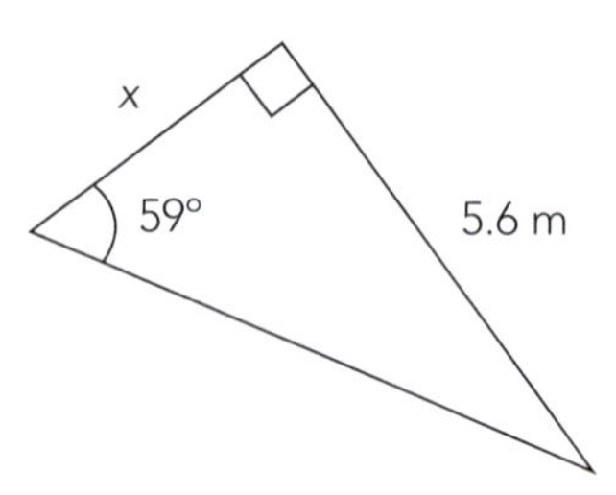

7

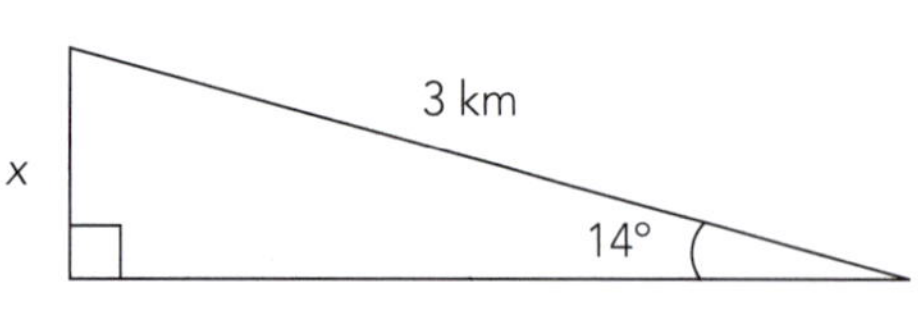

8

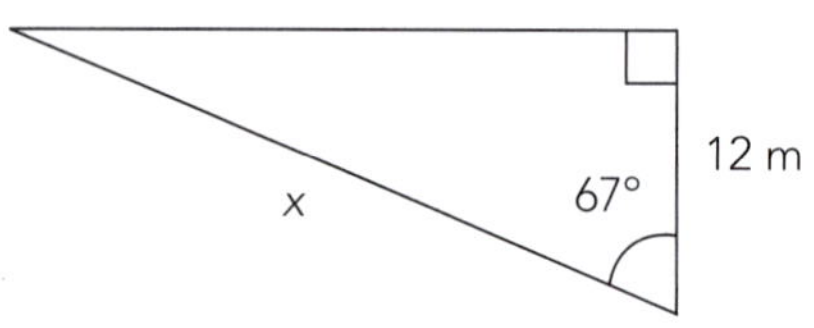

 ISBN: 9780170371629

9

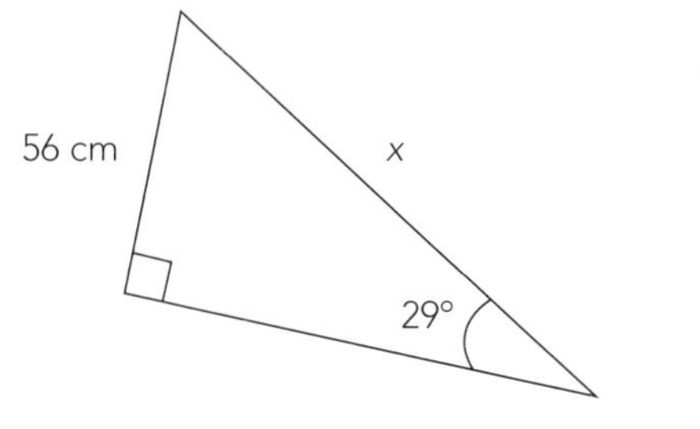

10

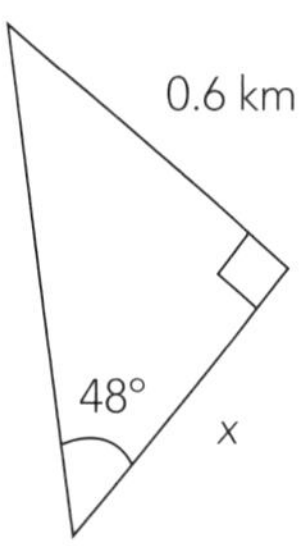

11

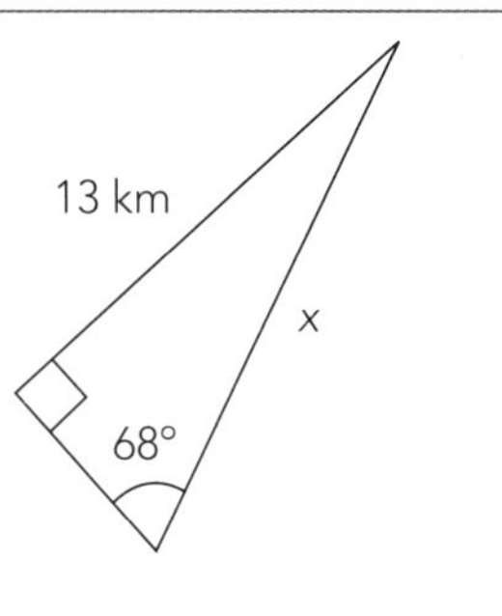

12

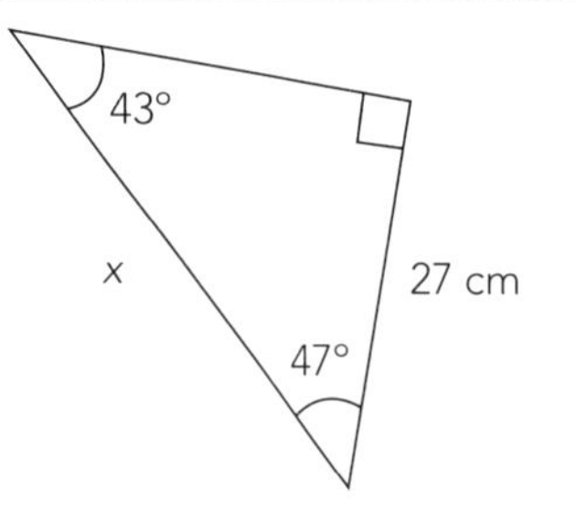

13

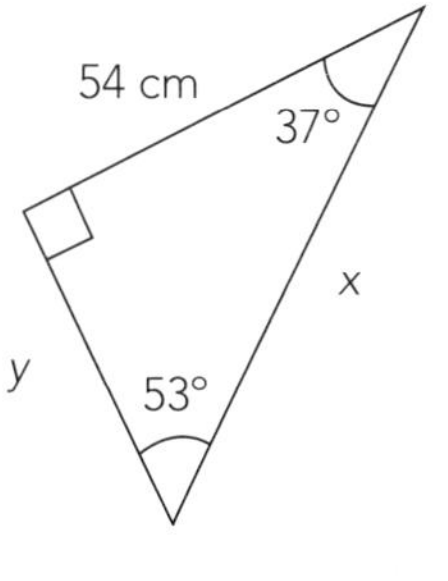

14

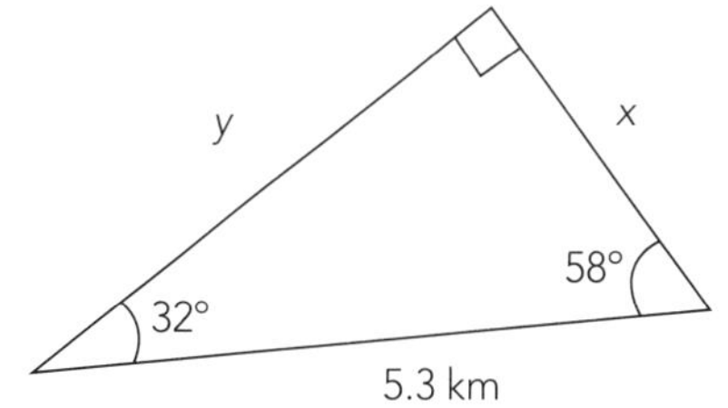

15

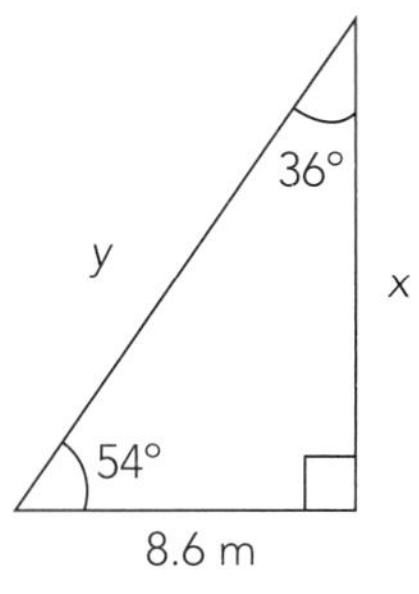

16

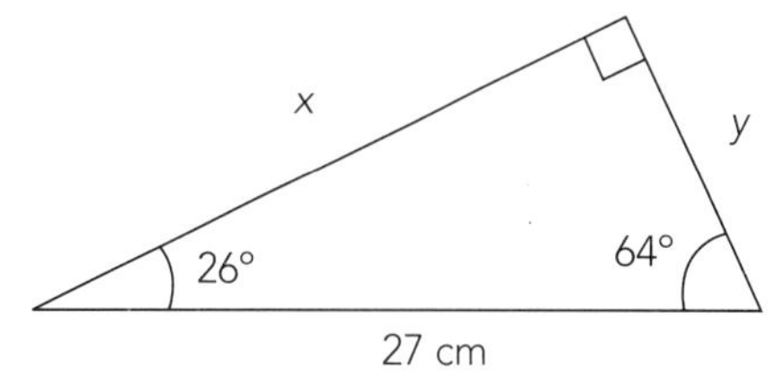

ISBN: 9780170371629

Finding sides — applications

1 A ladder is leaning against a wall. It makes an angle of 62° with the ground, and it reaches 2.5 m up the wall. How far (x) from the wall is the base of the ladder?

2.5 m

62°

x

2 A conveyor belt is required to take material from the ground floor of a factory to the first floor. The height difference is 3.2 m, and the maximum angle for moving the material is 25°. How long will the conveyor belt need to be, and what horizontal distance will be required?

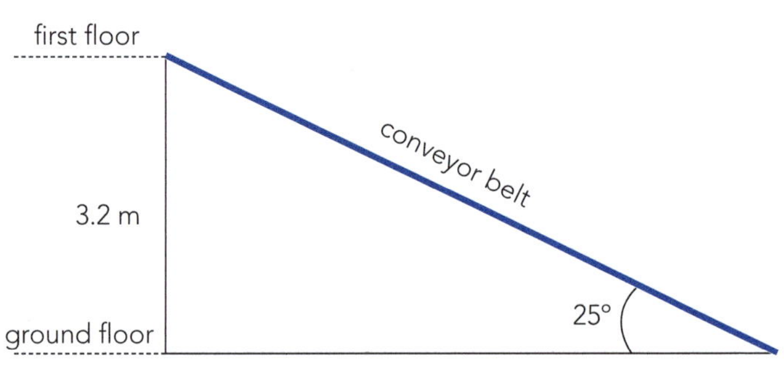

3 A wheelchair ramp is being built for a classroom. The angle between the horizontal and the ramp will be 4.8°. The difference between the height of the path and the classroom floor is 28 cm. How far from the classroom will the ramp need to start, and how long will it be?

classroom

28 cm

path

4.8°

ISBN: 9780170371629

4 In order to shed snow, the pitch on a roof must be at least 22.6°. If the roof must be 6 m wide, calculate its minimum height (h). How long will the roof be?

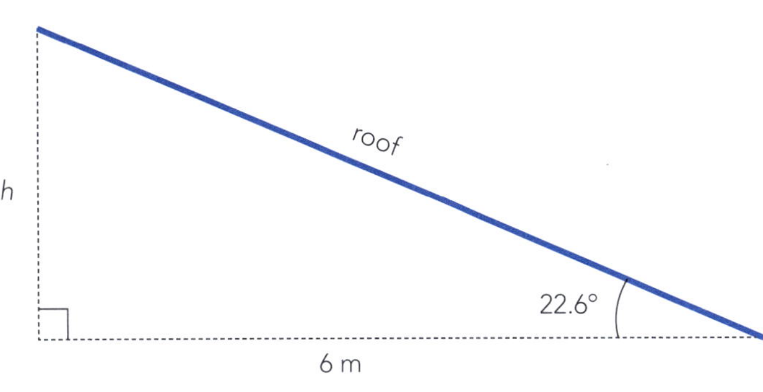

5 Kiri needs to know the height of a tree. From eye level, the angle between the horizontal and the top of the tree is 19° when she stands 11 m from the tree. Her eye is 1.52 m high. How high is the tree?

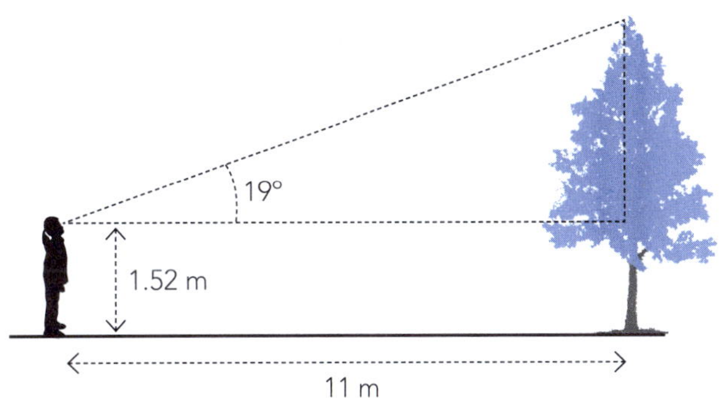

6 Two ski lifts take skiers to the top of a ski field. One is 540 m long at an angle of 22° to the horizontal. The second is 460 m long, and at an angle of 41°.
Find h, the difference in height between the bottom and the top of the field.

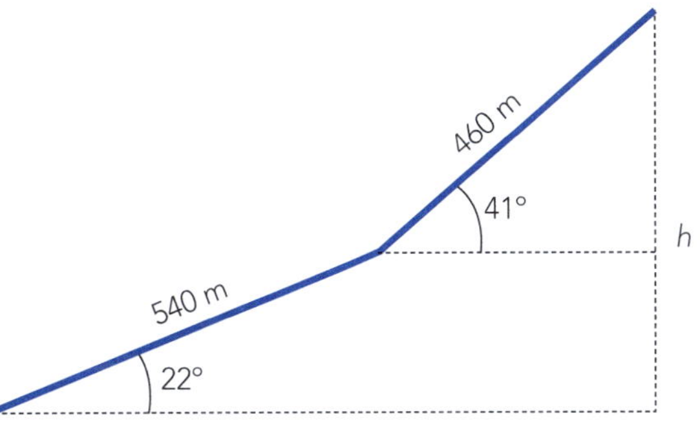

Finding angles using the sine

Once again we use the rule: $\sin \theta = \frac{O}{H}$

Example: Calculate the angle marked θ.

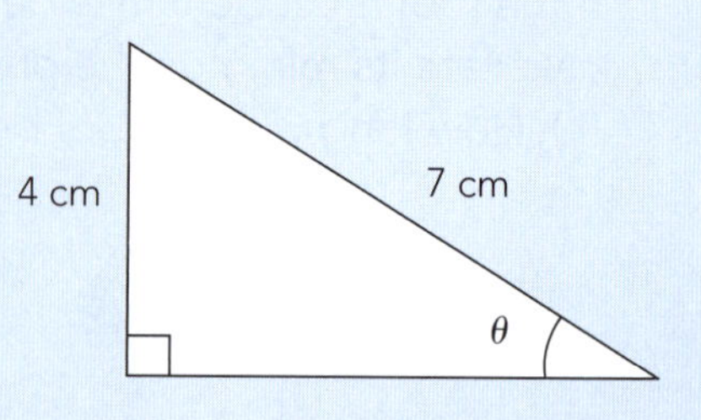

Step 1: **Label** the sides that are involved in the problem with 'A', 'O' and 'H'.

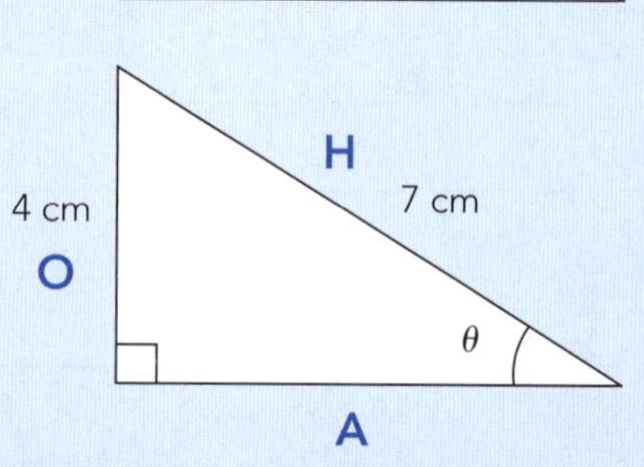

Step 2: Decide **which relationship** involves these sides. In this case we are concerned only with O and H so we must use:

$$\sin \theta = \frac{O}{H}$$

Step 3: **Substitute** the values from the triangle.

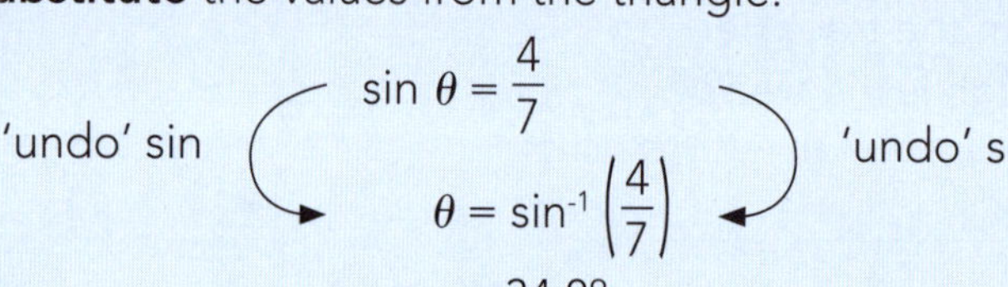

$$\sin \theta = \frac{4}{7}$$

$$\theta = \sin^{-1}\left(\frac{4}{7}\right)$$

$$= 34.8°$$

Use the 'inverse sin' button on your calculator. Don't forget the **brackets**.

Step 4: **Think about your answer — does it seem reasonable?**
In this case, an answer of 34.8° is reasonable because it is less than 90°.

Find the missing angles of these triangles.

1

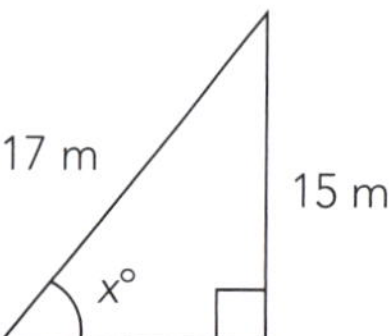

____

____

____

2

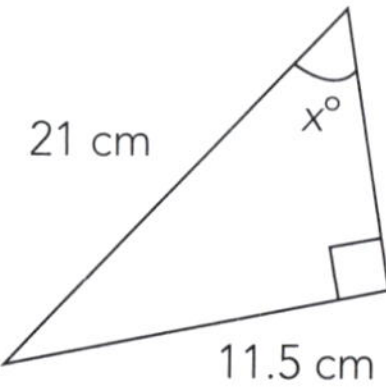

____

____

____

3

748 mm, 456 mm, x°

____

____

____

4

x°, 2.92 km, 5.7 km

____

____

____

 ISBN: 9780170371629

5

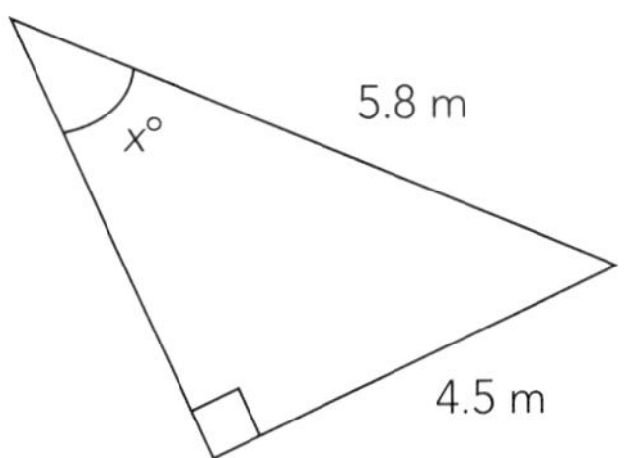

6

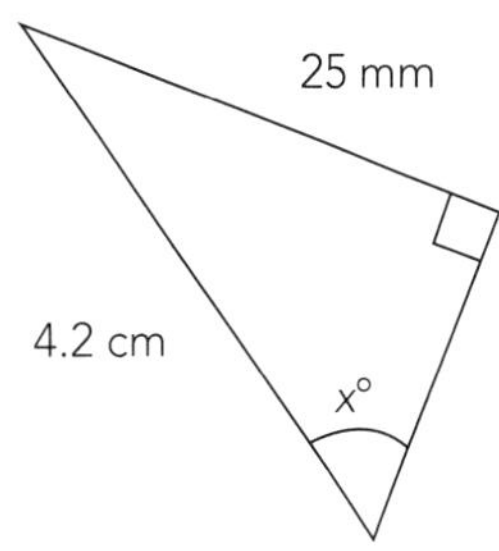

7

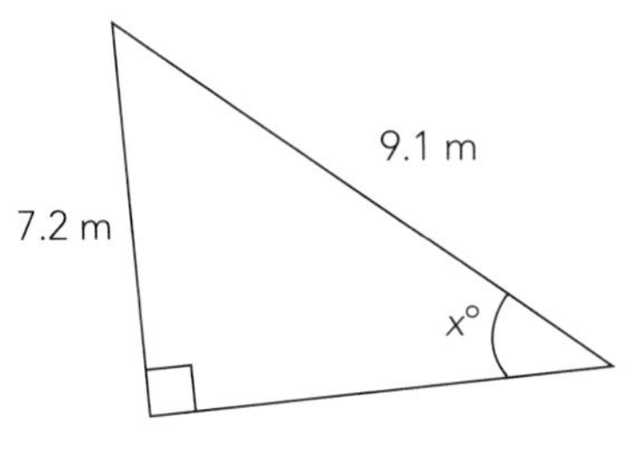

8

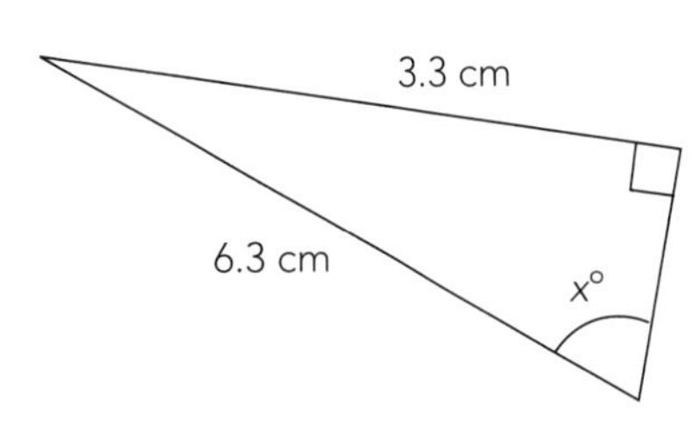

9

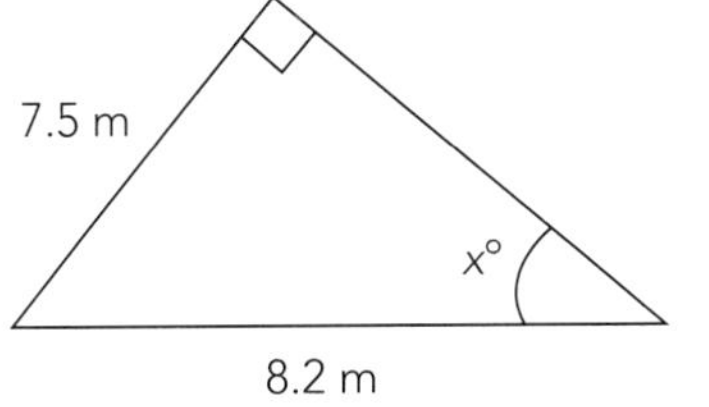

10

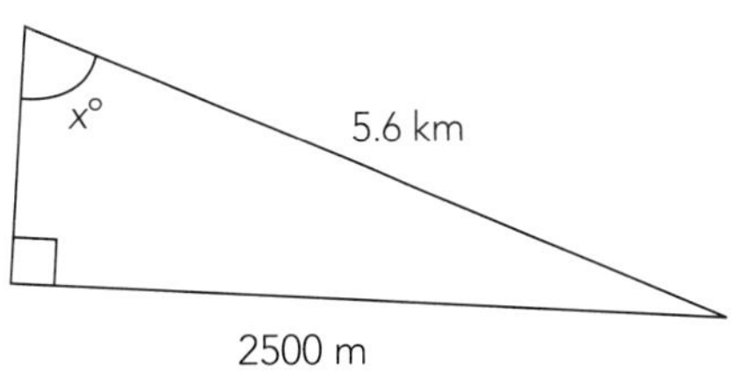

11

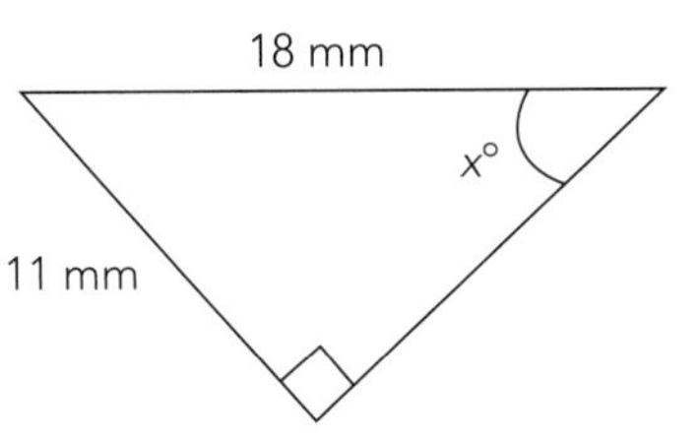

12

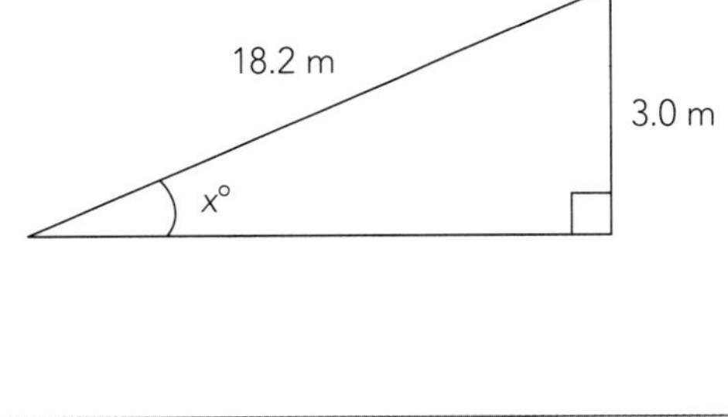

ISBN: 9780170371629

Finding angles using the cosine

Example: Calculate the angle marked θ.

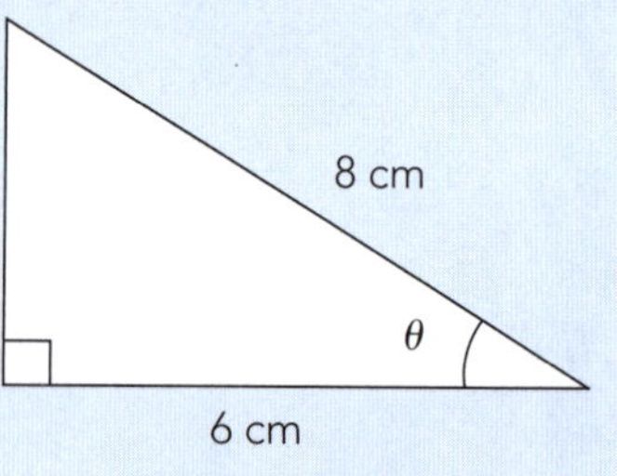

Step 1: **Label** the sides that are involved in the problem with 'A', 'O' and 'H'.

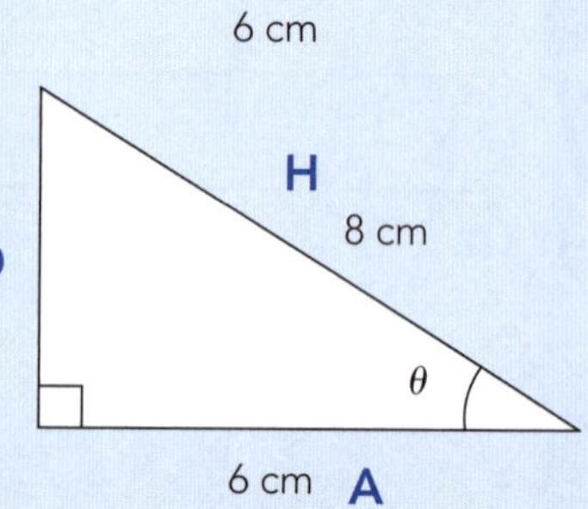

Step 2: Decide **which relationship** involves these sides. In this case we are concerned only with O and H so we must use:

$$\cos\theta = \frac{A}{H}$$

Step 3: **Substitute** the values from the triangle.

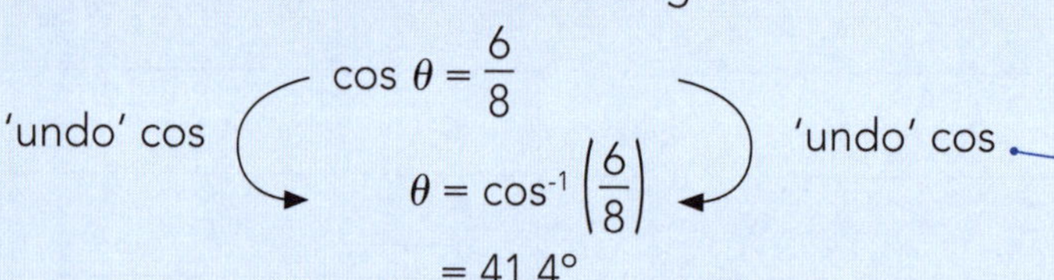

$$\cos\theta = \frac{6}{8}$$

$$\theta = \cos^{-1}\left(\frac{6}{8}\right)$$

$$= 41.4°$$

Use the 'inverse cos' button on your calculator. Don't forget the **brackets**.

Step 4: **Think about your answer — does it seem reasonable?**
In this case, an answer of 41.4° is reasonable because it is less than 90°.

Find the missing angles of these triangles.

1

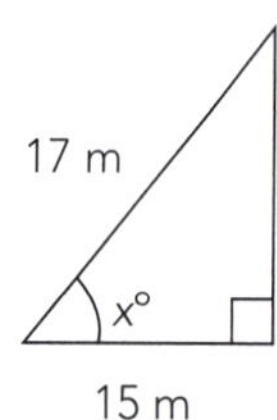

2

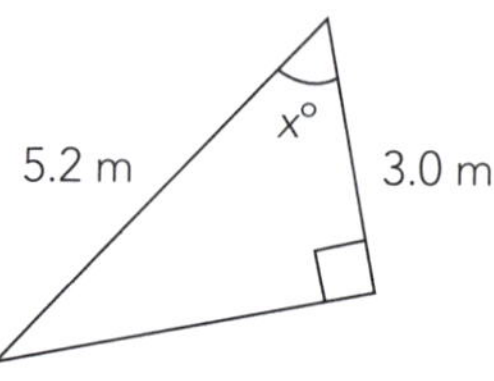

3

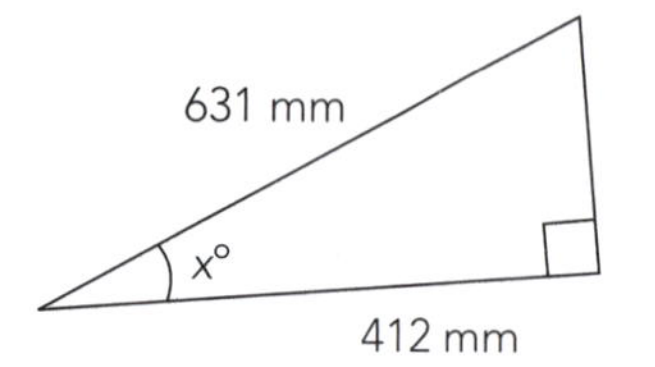

4

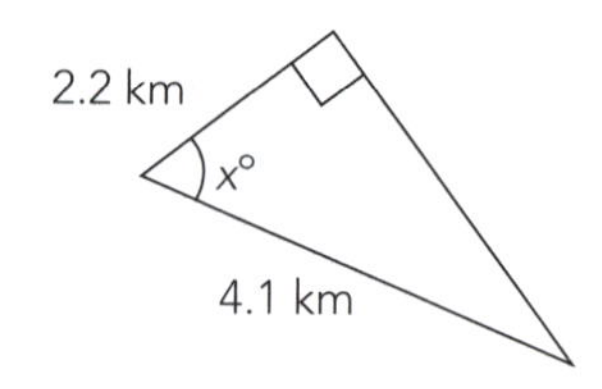

ISBN: 9780170371629

5

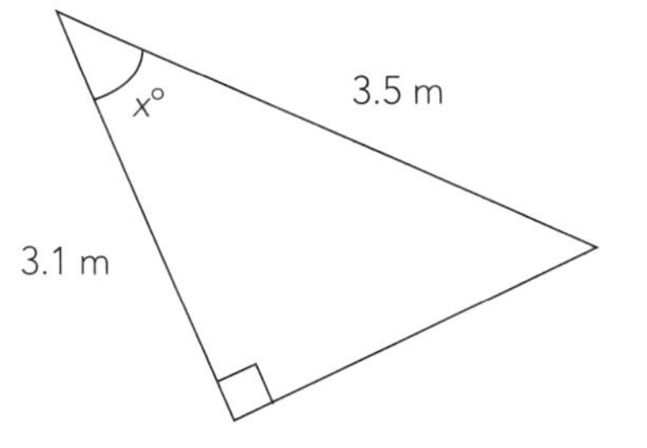

6

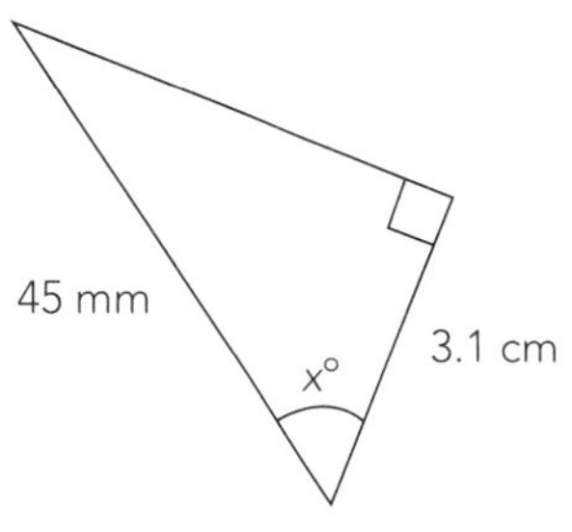

7

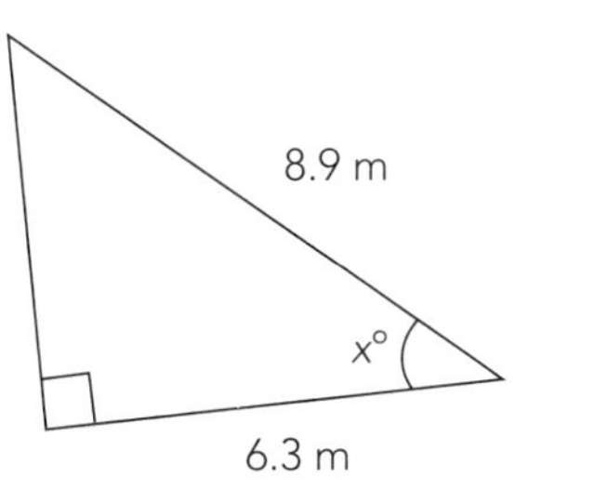

8

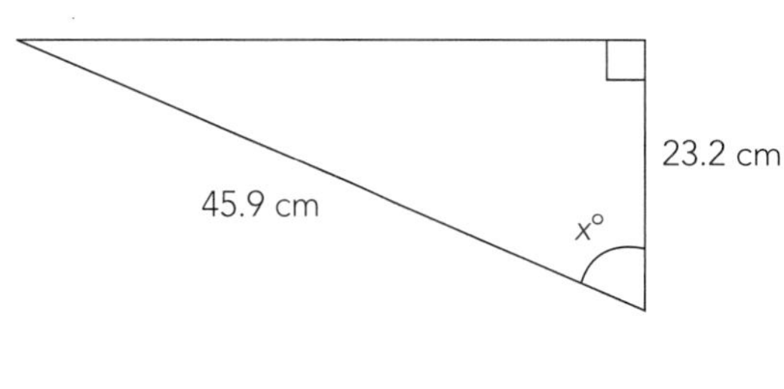

9

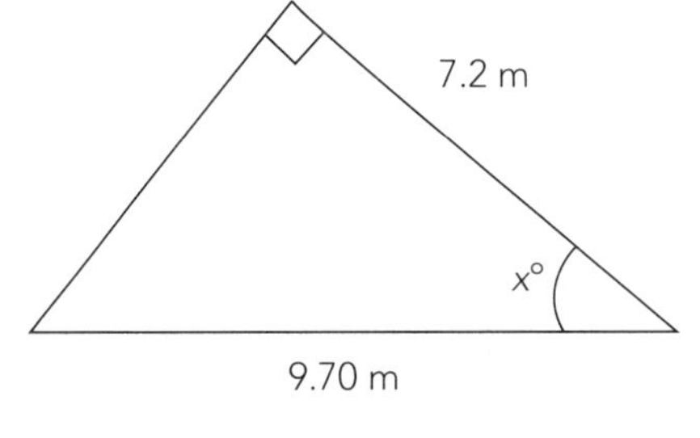

10

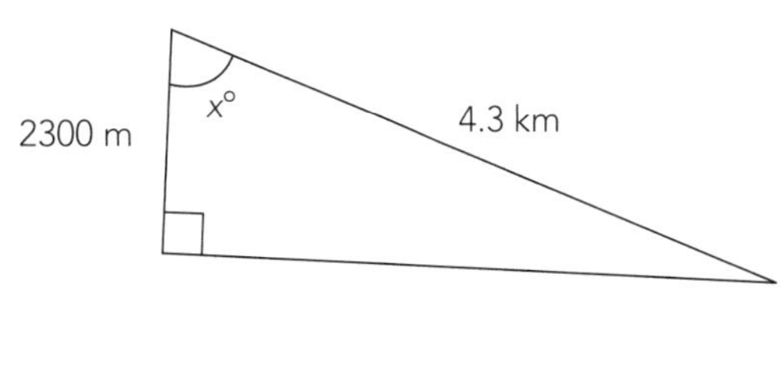

11

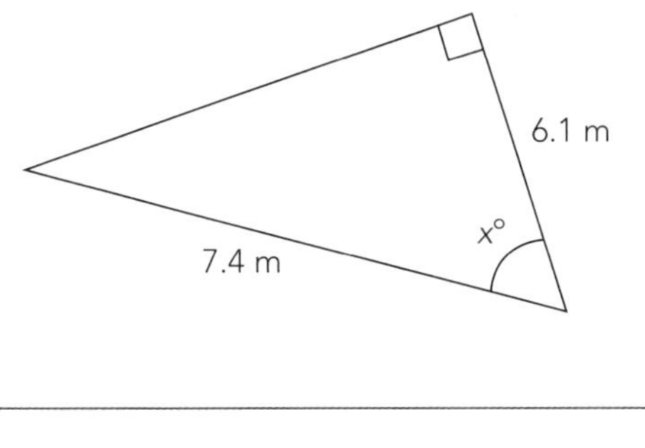

12

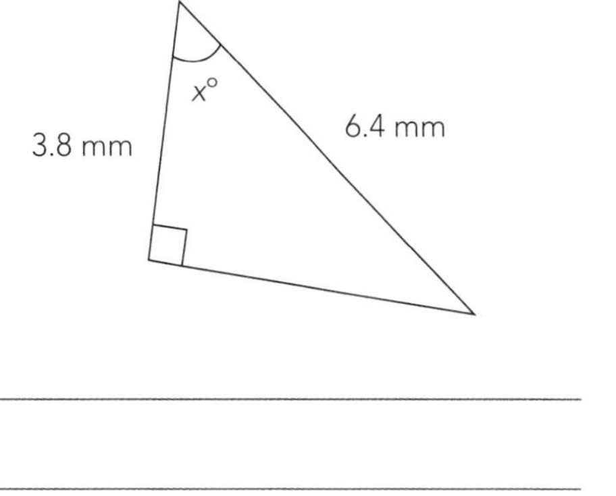

ISBN: 9780170371629

Finding angles using the tangent

Example: Calculate the angle marked θ.

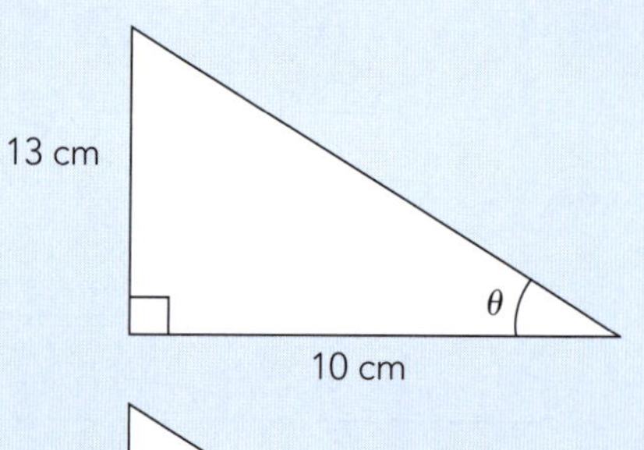

Step 1: **Label** the sides that are involved in the problem with 'A', 'O' and 'H'.

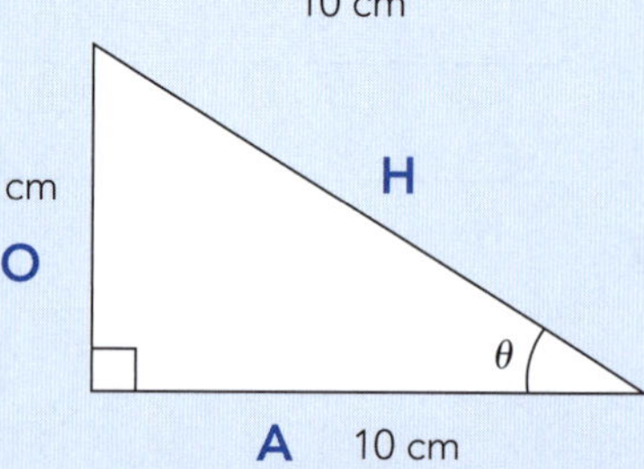

Step 2: Decide **which relationship** involves these sides. In this case we are concerned only with O and A so we must use:

$$\tan\theta = \frac{O}{A}$$

Step 3: **Substitute** the values from the triangle.

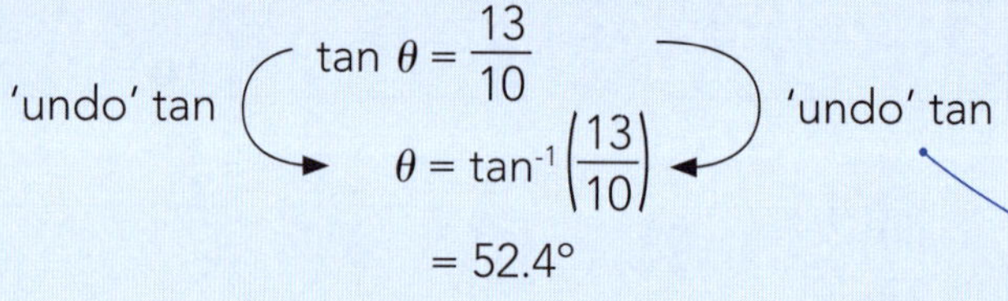

$$\tan\theta = \frac{13}{10}$$

$$\theta = \tan^{-1}\left(\frac{13}{10}\right)$$

$$= 52.4°$$

Use the 'inverse tan' button on your calculator. Don't forget the **brackets**.

Step 4: **Think about your answer — does it seem reasonable?**
In this case, θ must be more than 45° because 13 cm is longer than 10 cm, so an answer of 52.4° is reasonable.

Find the missing angles of these triangles.

1

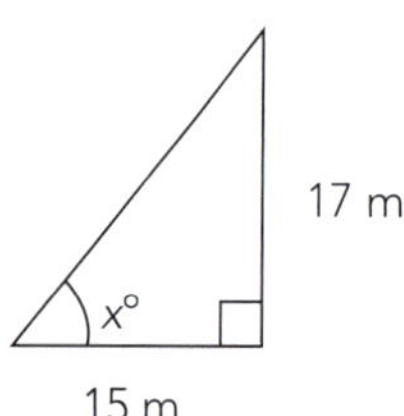

2

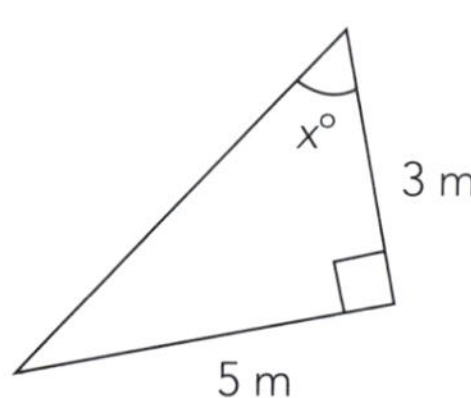

3

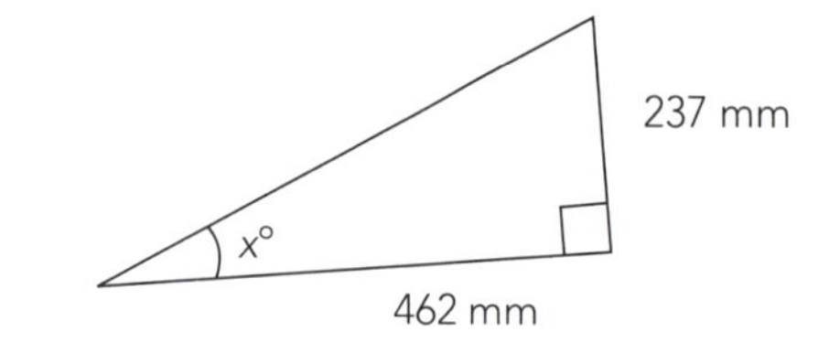

4

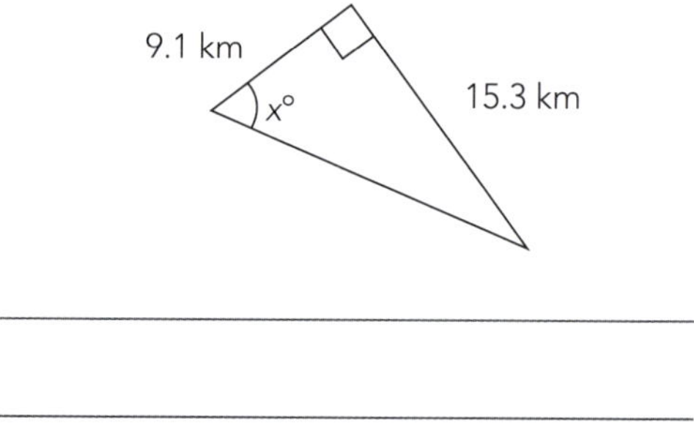

ISBN: 9780170371629

5

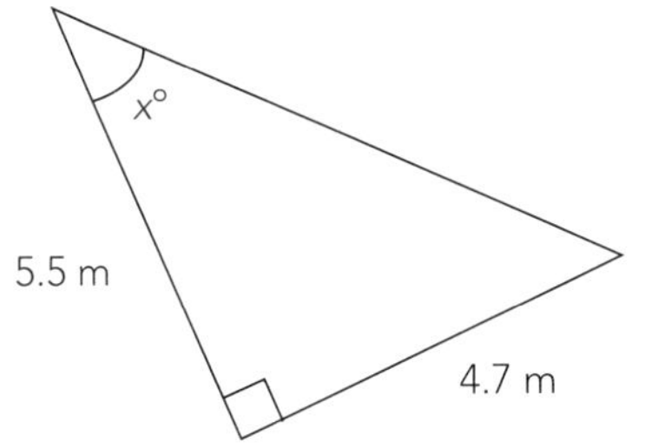

6

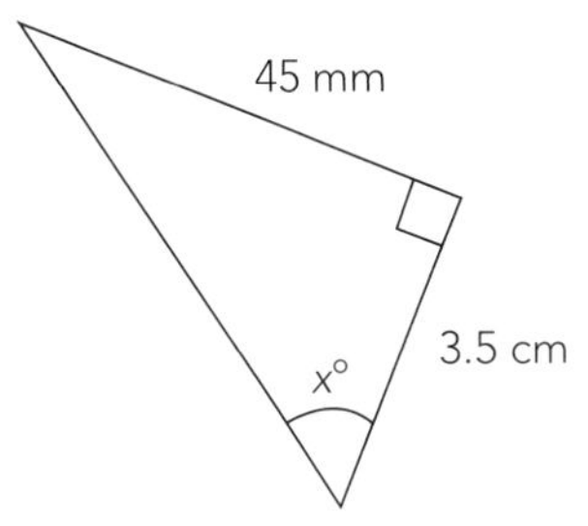

7

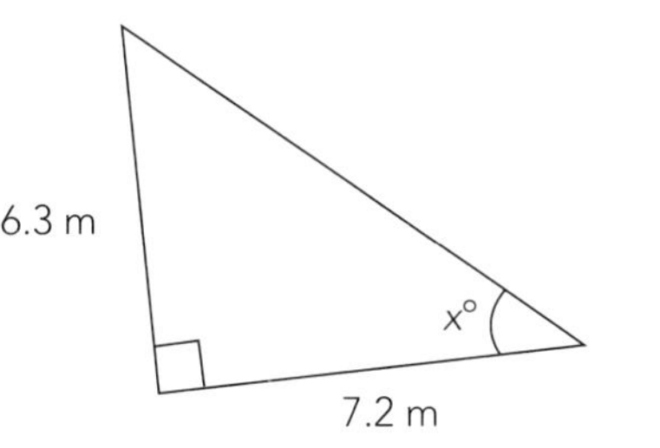

8

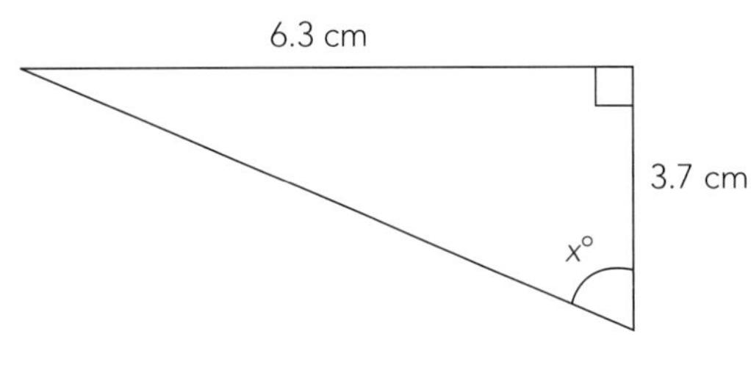

9

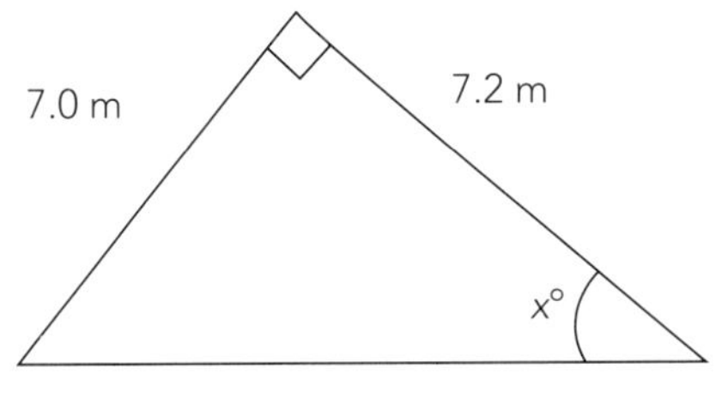

10

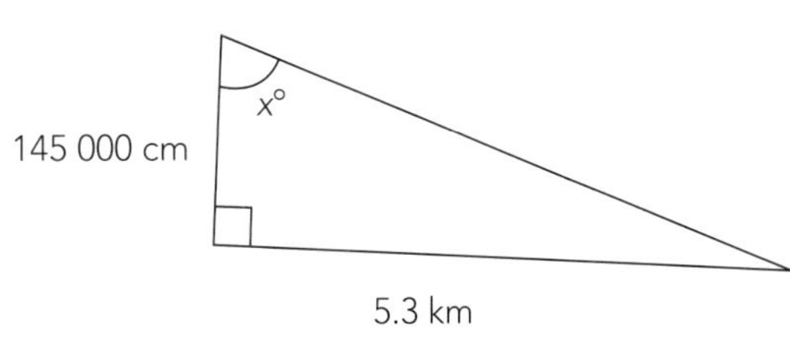

11

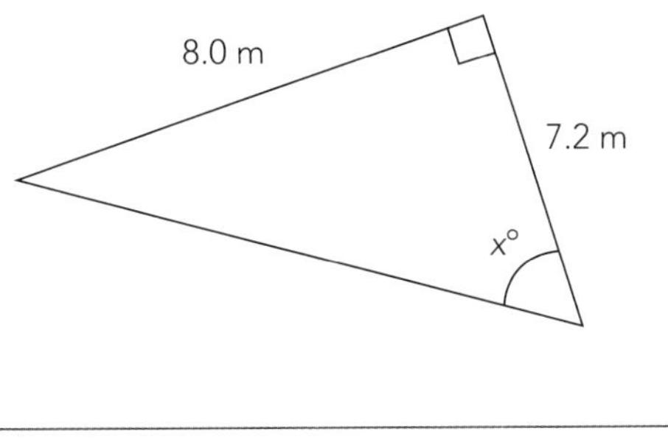

12

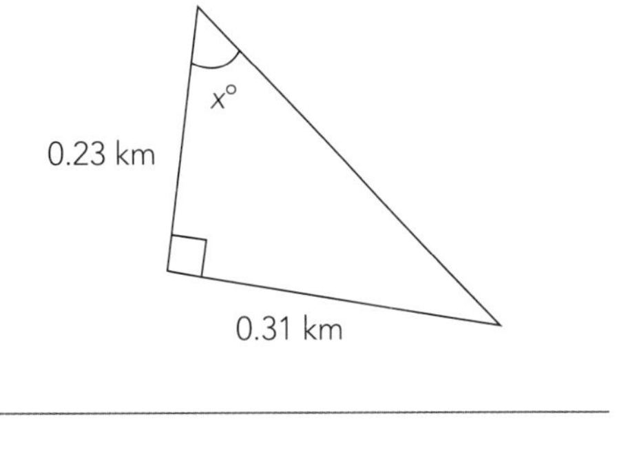

ISBN: 9780170371629

Finding angles — mixing it up

Find the missing angles of these triangles.

1

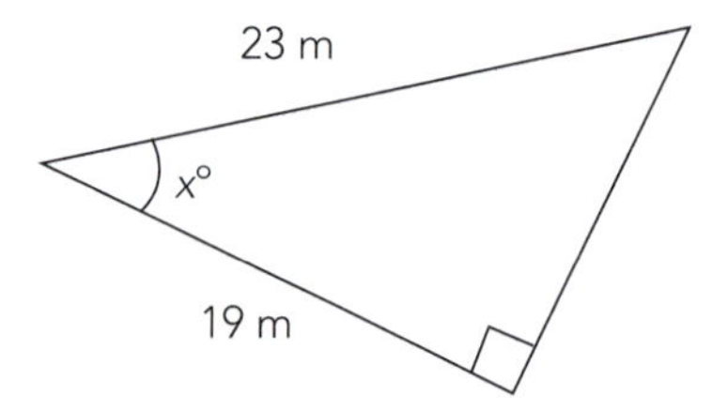

2

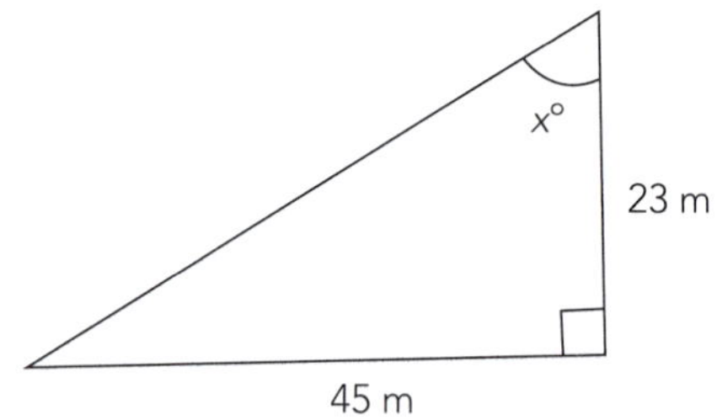

3

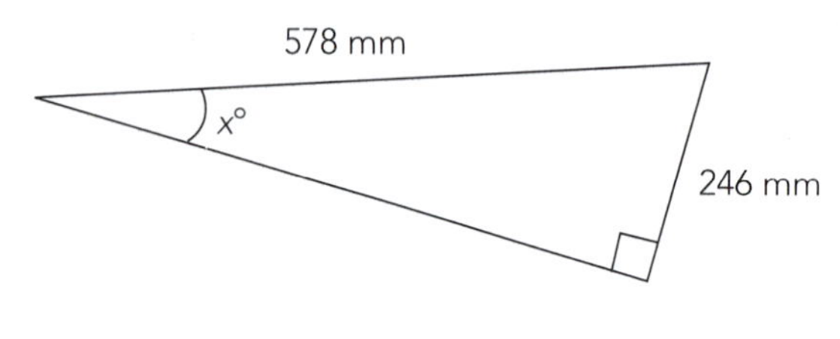

4

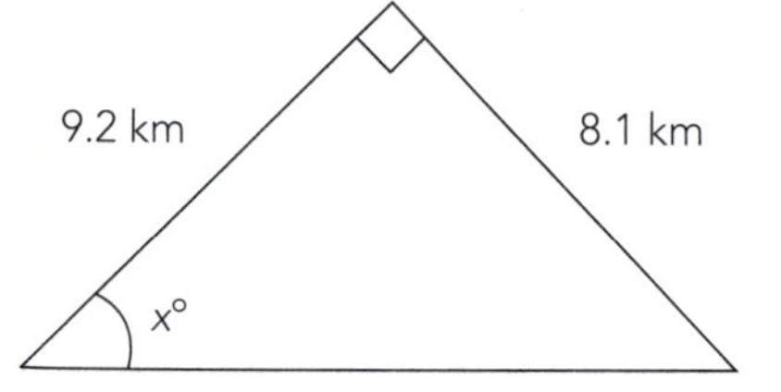

5

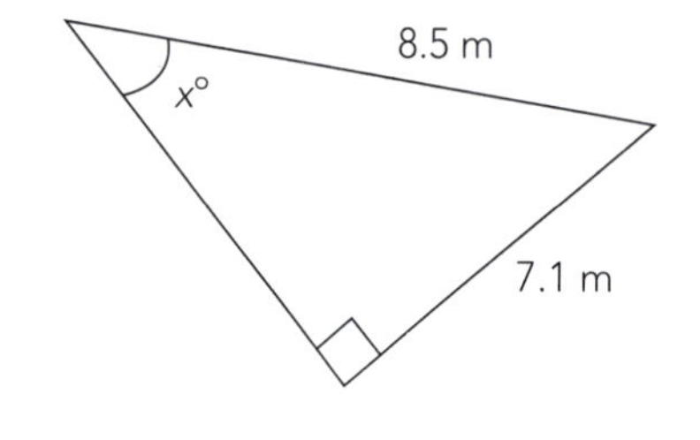

6

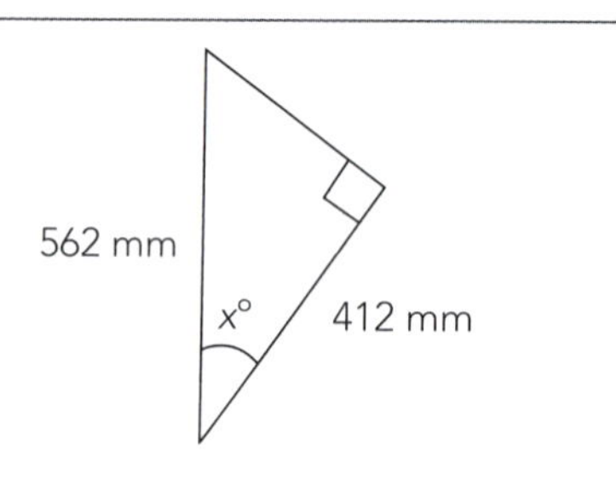

7

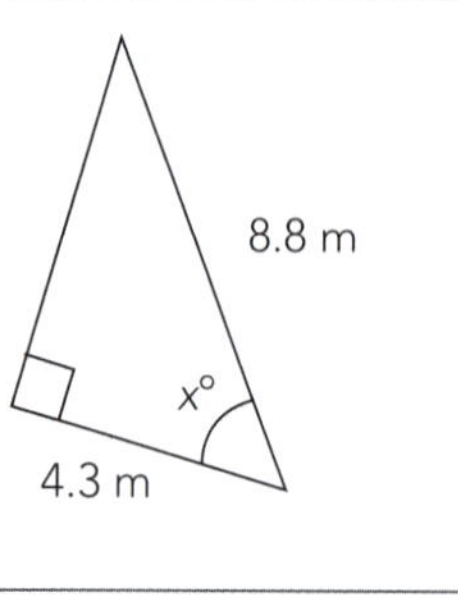

8

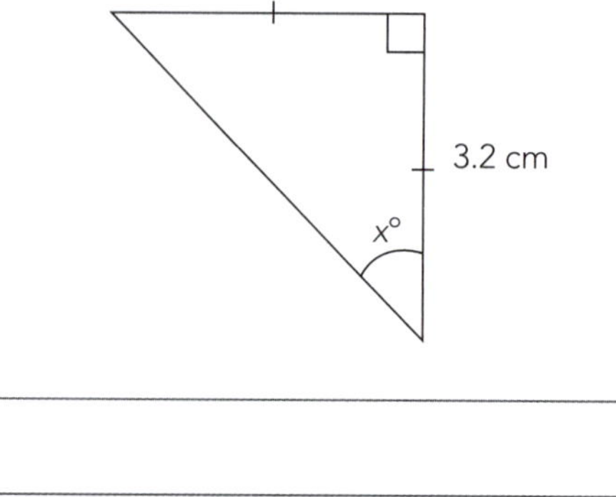

ISBN: 9780170371629

9

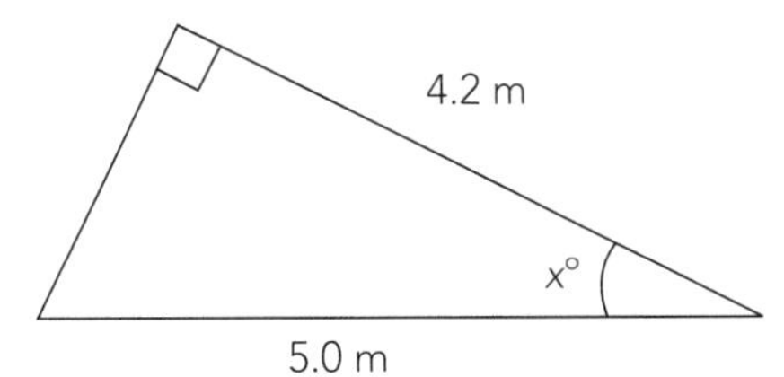

10

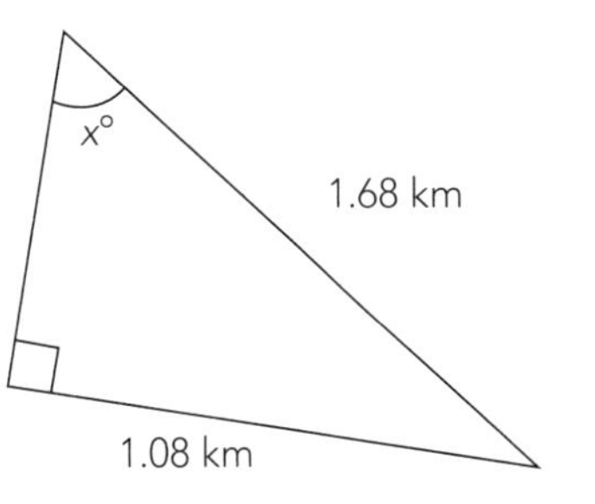

11

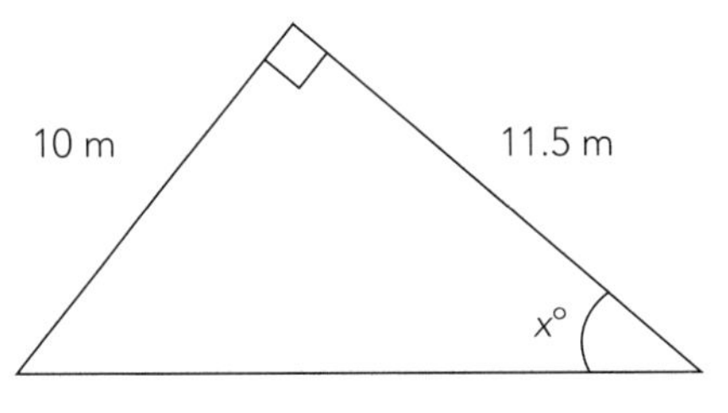

12

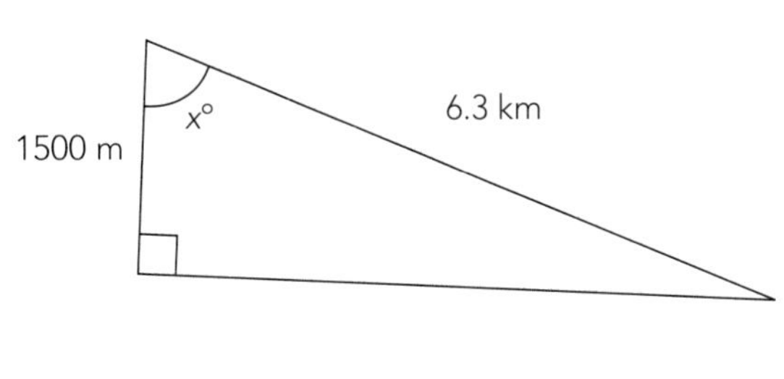

13

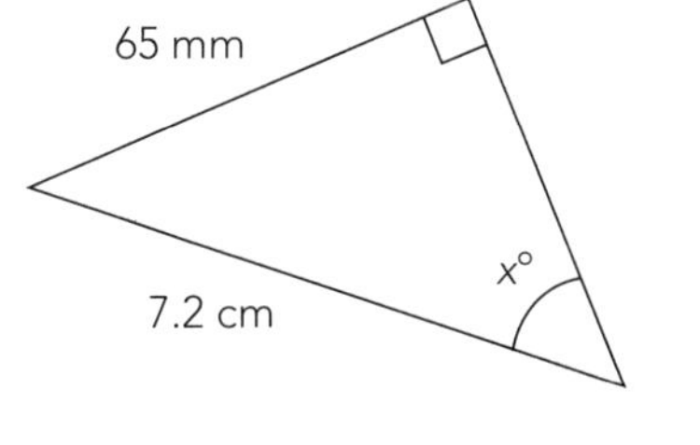

14

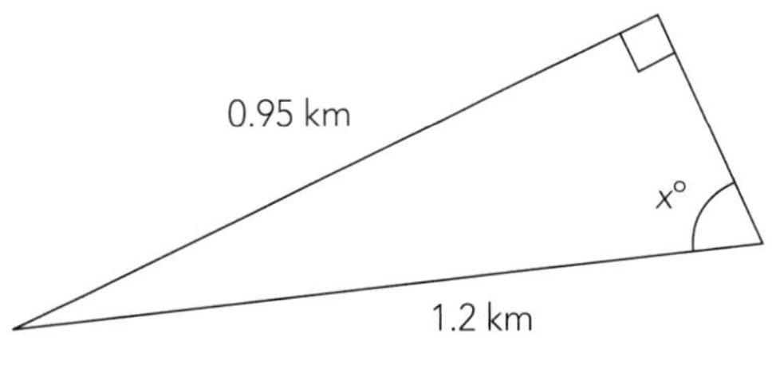

15

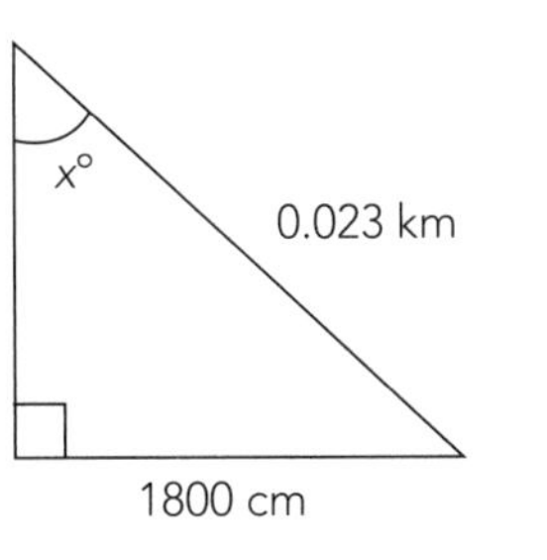

16

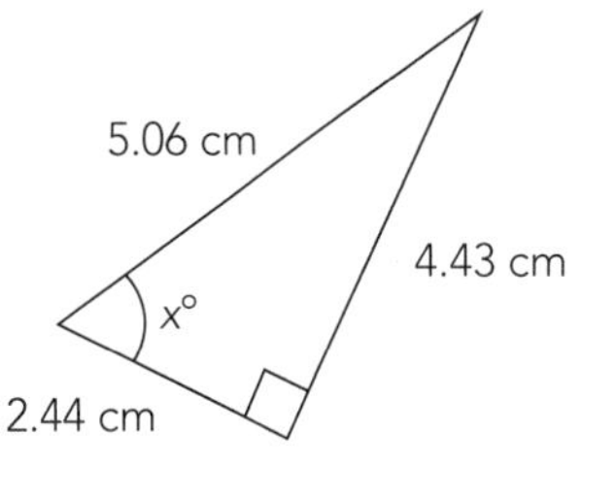

ISBN: 9780170371629

Finding angles — applications

1 A ladder is leaning against a wall. It reaches 3 m up the wall, and its base is 1.44 m from the base of the wall. Calculate the angle (θ) the ladder makes with the ground.

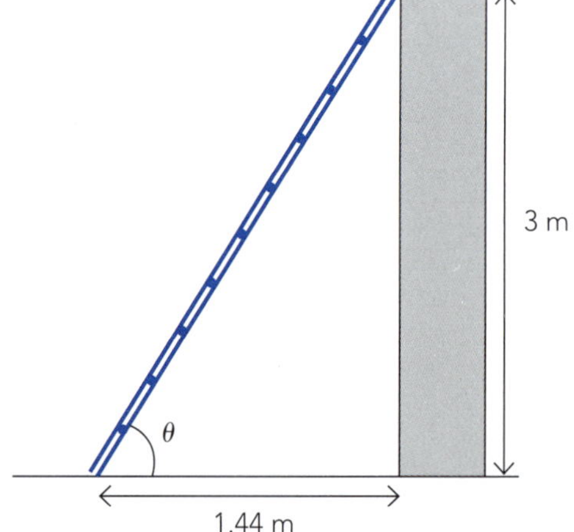

2 The shadow cast by a 6 m flag pole is 9 m long. Calculate the angle (θ) that the sun's rays make with the ground.

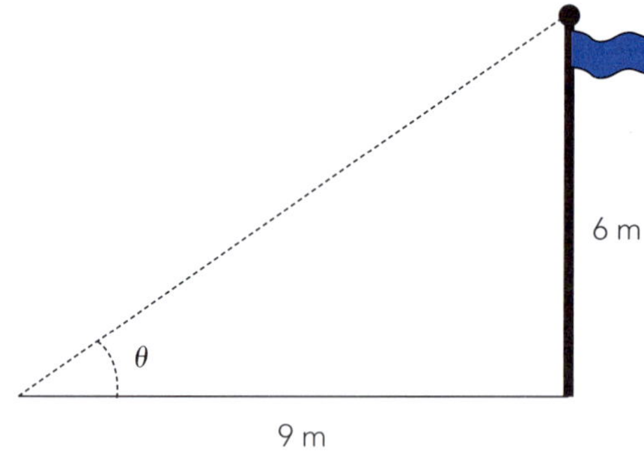

3 Hare needs to know the height of a tree. He holds a 2 m post vertically and measures the length of its shadow (1.739 m). He then measures the length of the tree's shadow (5.363 m). Calculate the height of the tree.

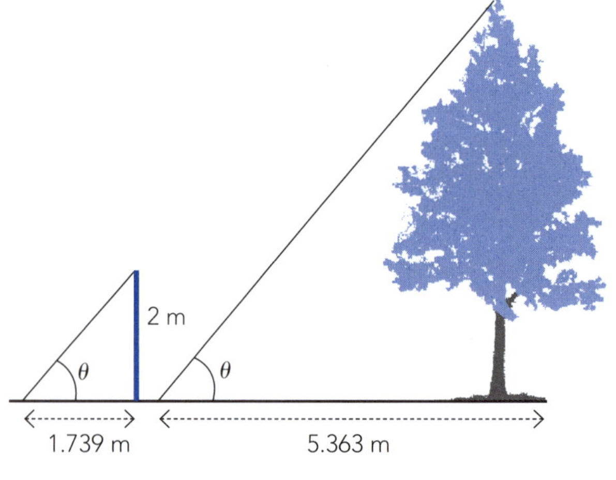

ISBN: 9780170371629

4 To get the ideal pitch on a roof for photo-voltaic cells, it needs to go up 2 units for every 3 units across. Calculate the angle the roof makes with the horizontal.

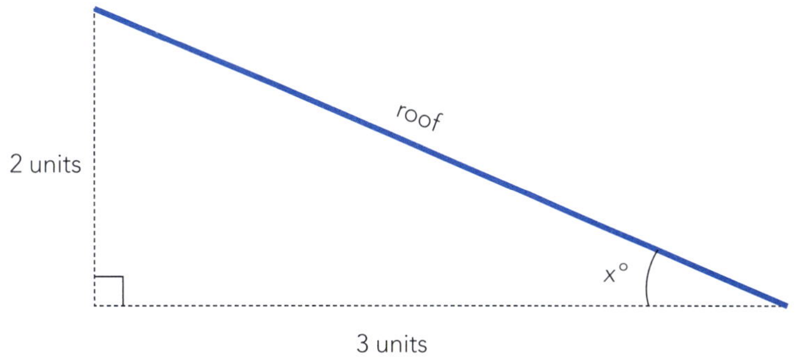

roof

2 units

$x°$

3 units

5 A travellator is to be installed in a new shopping mall. It needs to lift people exactly 3 m between the first floor and the car park. The maximum length the travellator can be is 15 m. Building regulations state that the maximum angle for a travellator is 12°. Will this new travellator comply with the regulations?

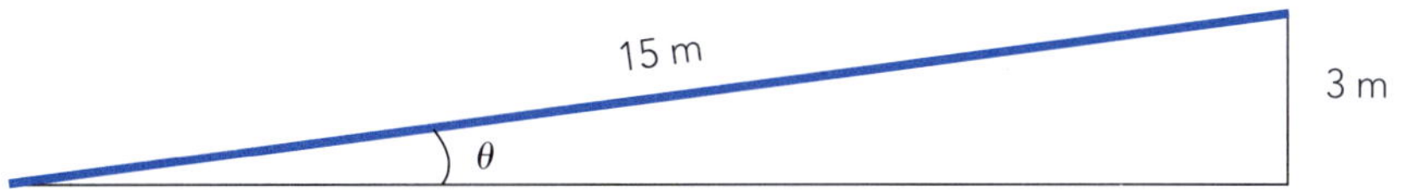

15 m

3 m

θ

6 A wheelchair ramp is being built for a classroom. The building code states that angle between the horizontal and the ramp must not be more than 4.8°. The difference between the height of the path and the classroom floor is 34 cm. If the maximum horizontal distance for the ramp is 4 m, will the ramp comply with the building code?

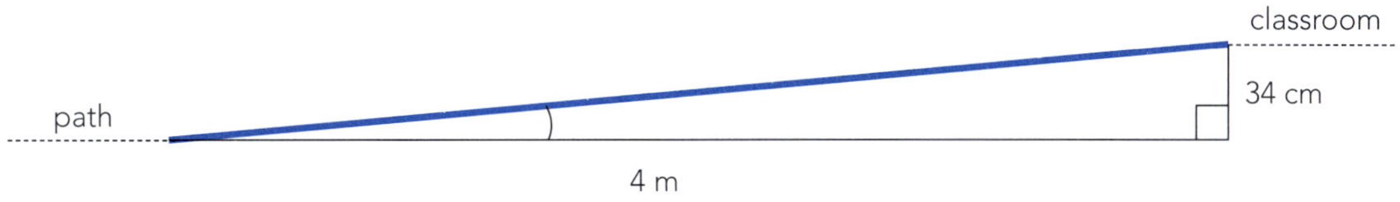

classroom

34 cm

path

4 m

Putting it together — with Pythagoras

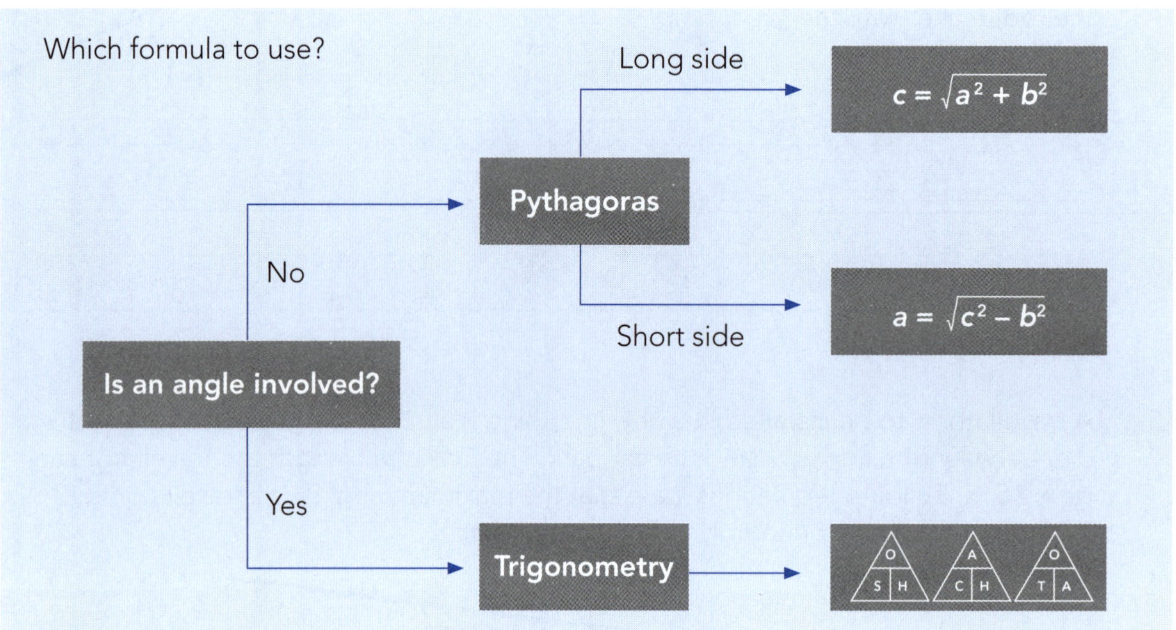

Calculate the unknown sides and angles.

1

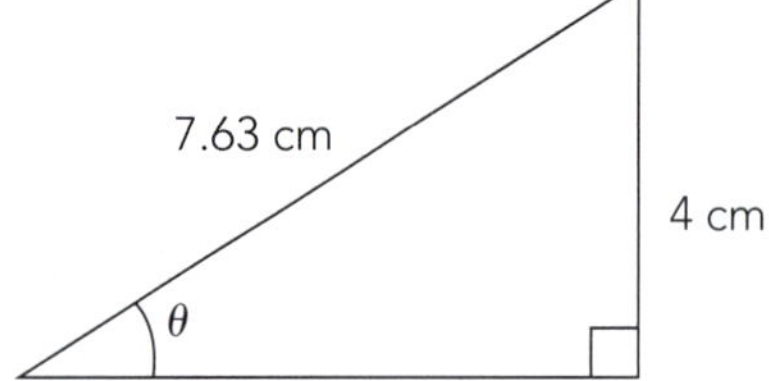

2

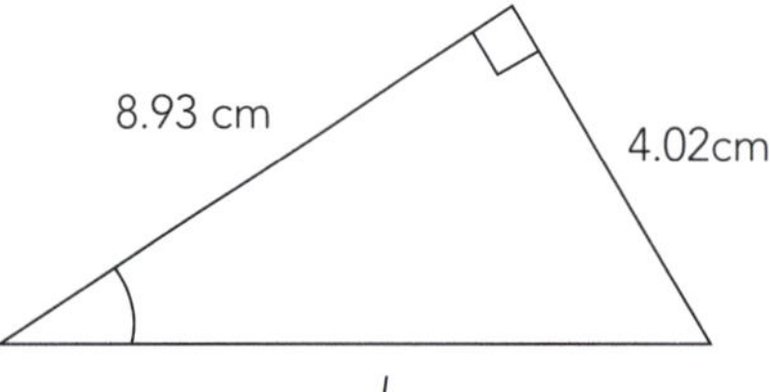

3

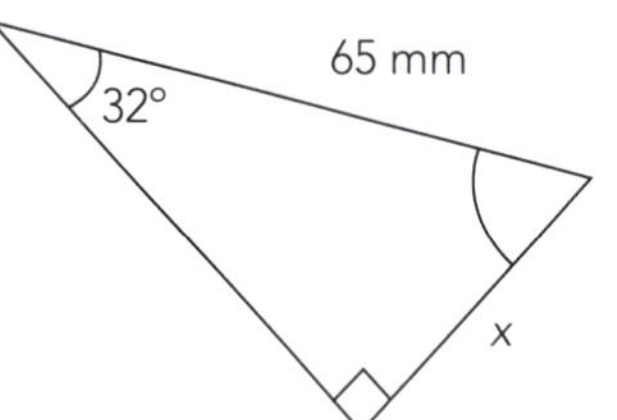

4

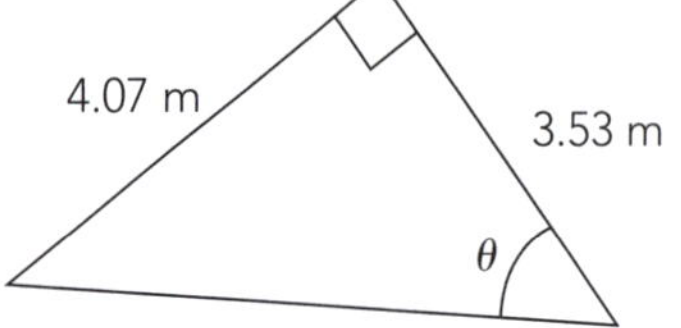

 ISBN: 9780170371629

5

3.62 km

D

6.45 km

6

66°

H

0.58 km

7

y

428 mm

25°

8

θ

1.08 m

0.88 m

9

64.1 cm

θ

78.4 cm

10 A rule for building ramps is that they may rise only 1 unit for every 12 units of horizontal length. Calculate:

a the angle between the ramp and the ground

b the length of the ramp.

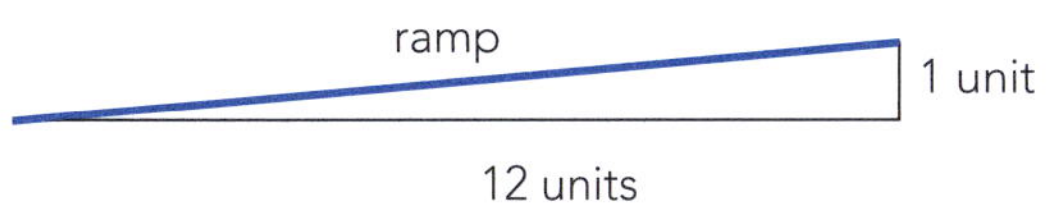

ISBN: 9780170371629

11 Calculate the angle between the ladder and the ground.

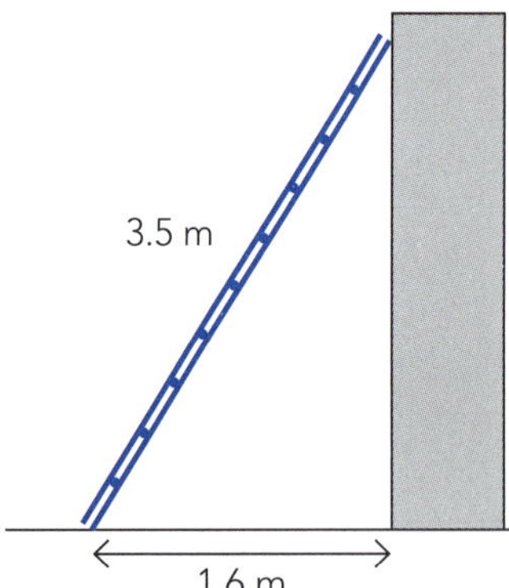

12 Calculate:

a the length L

b the angle θ.

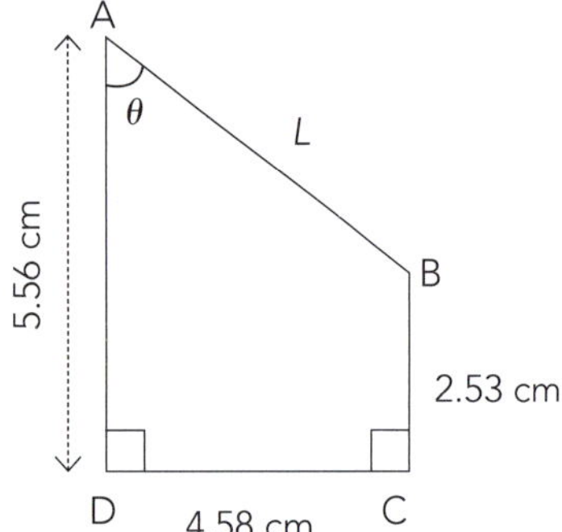

13 Calculate the lengths of:

a x

b y.

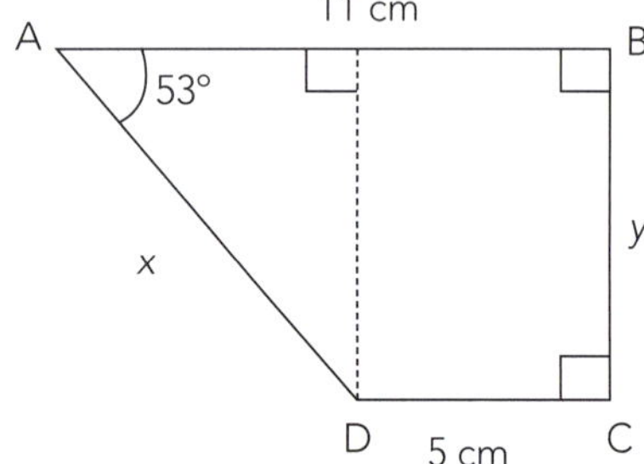

14 ABCD is a rhombus whose diagonals meet at E.

a Calculate θ.

b Calculate the length of diagonal AC.

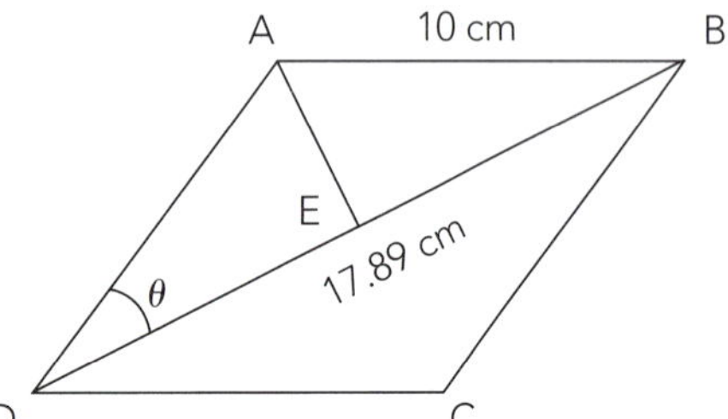

ISBN: 9780170371629

15 Calculate the length of AC.

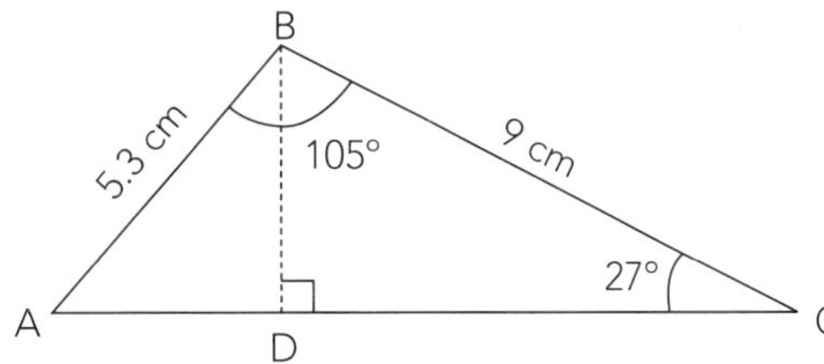

16 A 50 cm square has a triangle cut off it. The base of the triangle (XY) is also 50 cm. If AX = CY, calculate h, the height of the triangle.

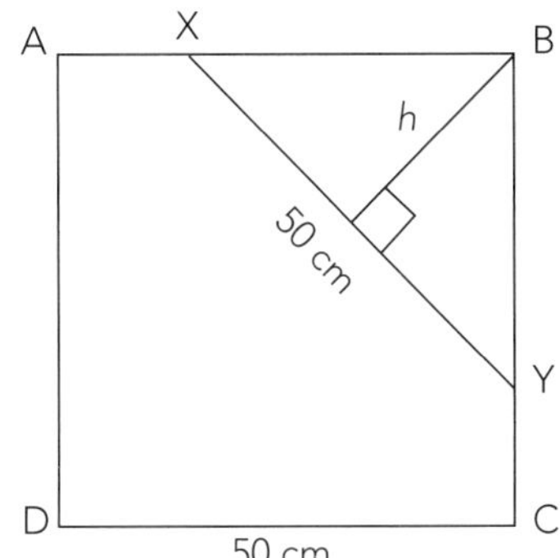

17 ADCE is a kite. The diagonal DE is 8 cm. Calculate the length of the diagonal AC.

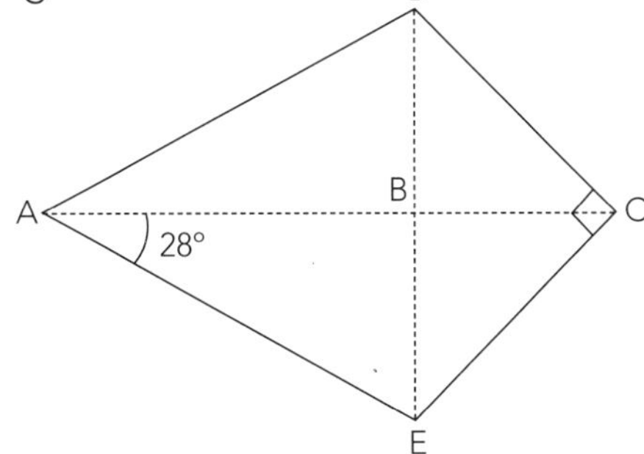

18 Calculate the perimeter of this figure.

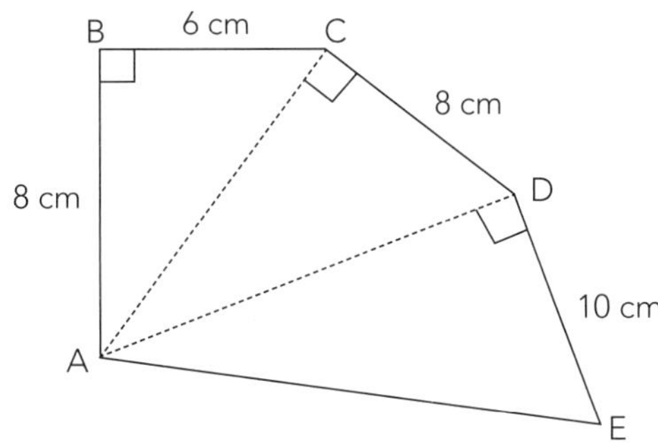

ISBN: 9780170371629

Similar triangles

- Similar triangles are the same shape.
- They have the same-sized angles as each other.

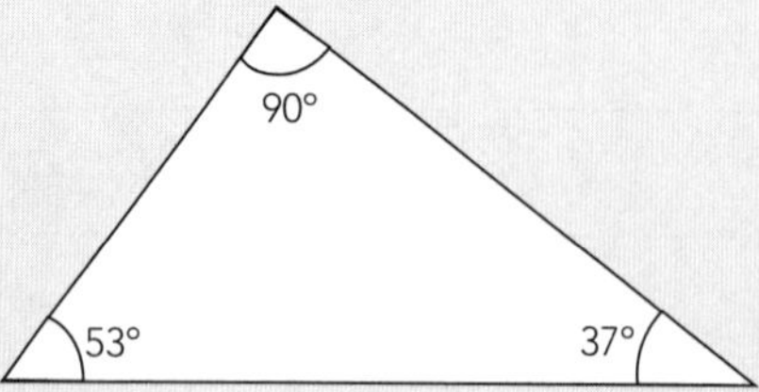

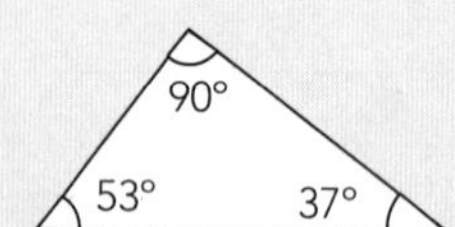

- They are not always drawn the same way up.

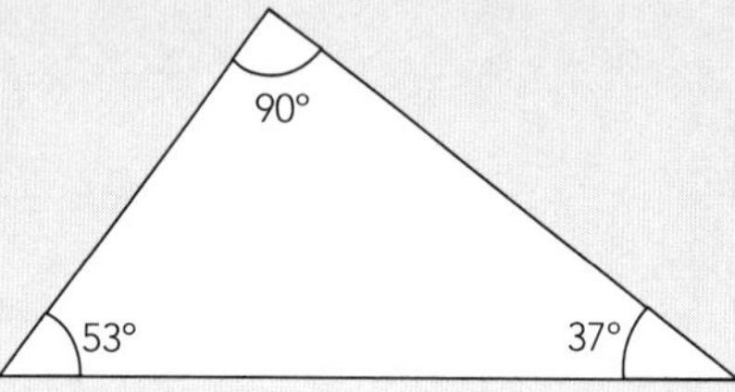

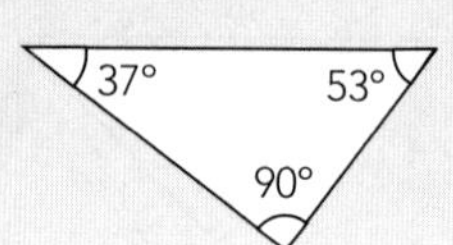

- The angles can also be in reverse order.

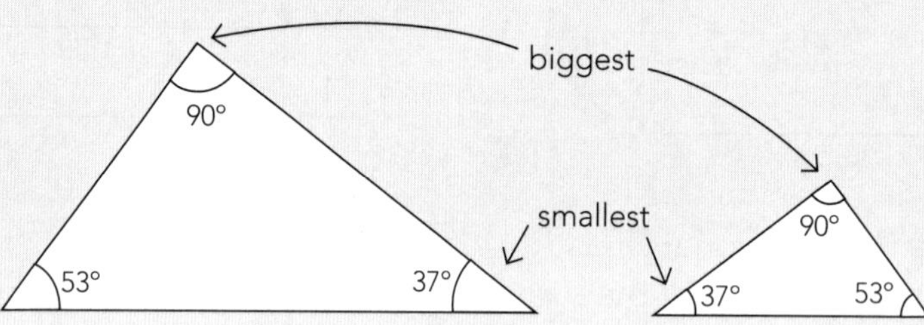

- The sides are also in proportion.

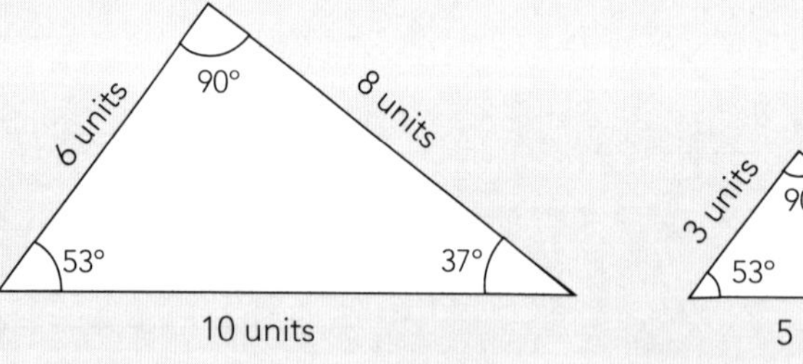

In this case: $\frac{\text{big triangle}}{\text{small triangle}} = \frac{10}{5} = \frac{8}{4} = \frac{6}{3} = 2$

- It is customary to label the vertices in corresponding order.

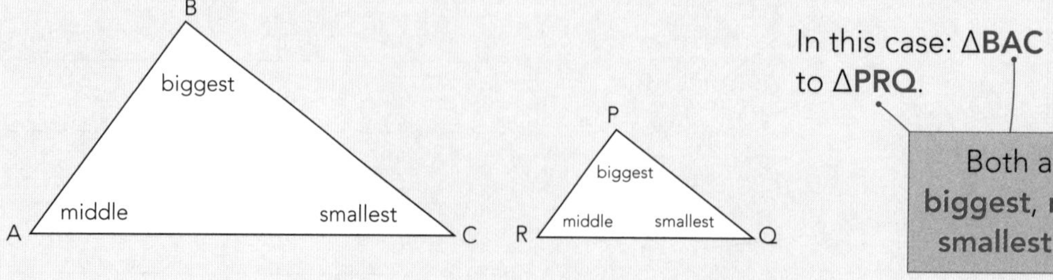

In this case: Δ**BAC** is similar to Δ**PRQ**.

Both are in **biggest, middle, smallest** order.

 ISBN: 9780170371629

Justification using angles

You must show that angles are equal.

Examples: State whether the following pairs of triangles are similar. Justify your answers.

1

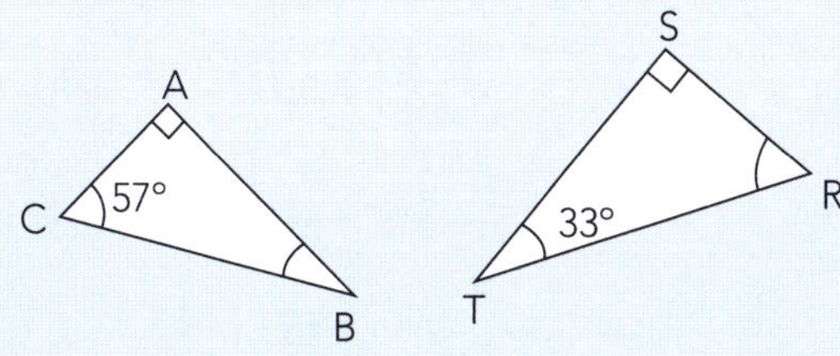

∠ABC = 33° (∠s in Δ = 180°)
∴ Δ**ABC** is similar to Δ**STR** because the angles are equal.

Biggest, smallest, middle.

2

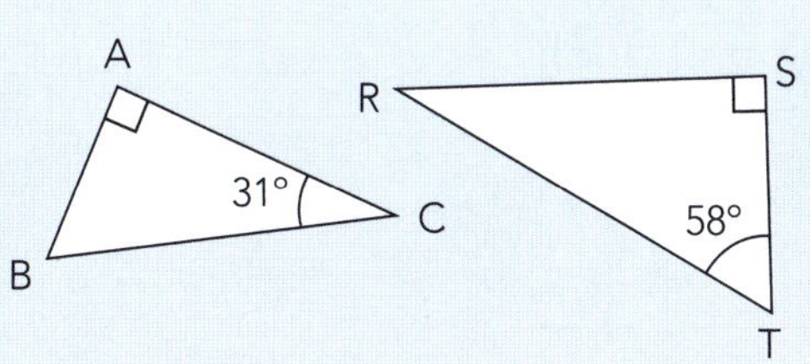

∠ABC = 59° (∠s in Δ = 180°)
∴ Δ**ABC** is not similar to Δ**RST** because the angles are not equal.

State whether the following pairs of triangles are similar or not, and justify your answer.

1

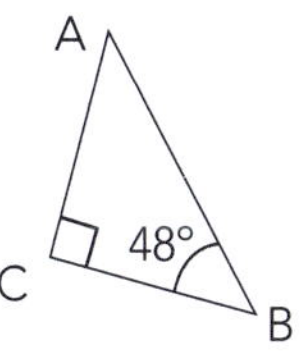

2

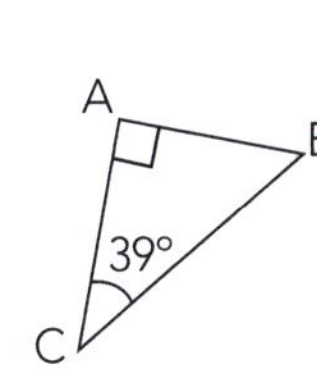

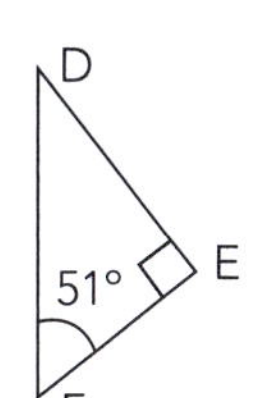

3

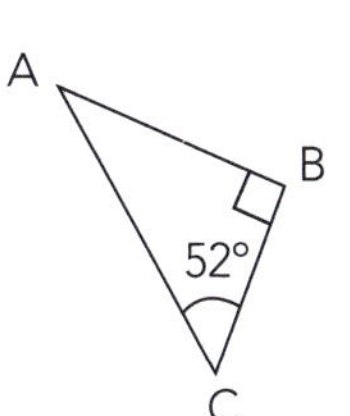

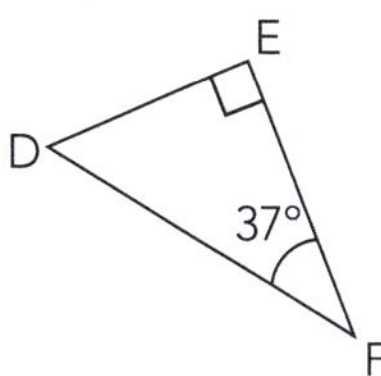

4

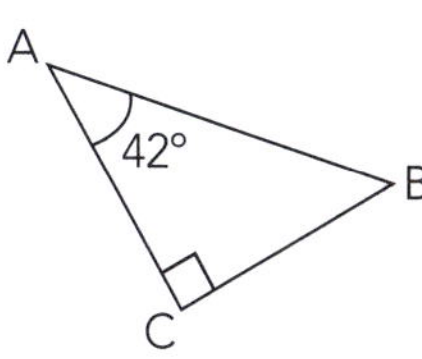

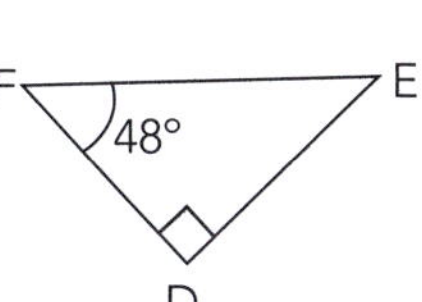

5

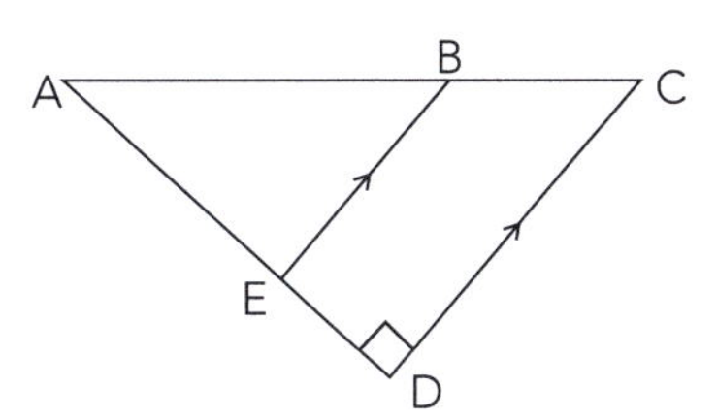

ISBN: 9780170371629

Justification using sides

You must show that the sides are in proportion.

Examples: Write a statement by each pair of triangles. Justify your answers.

1

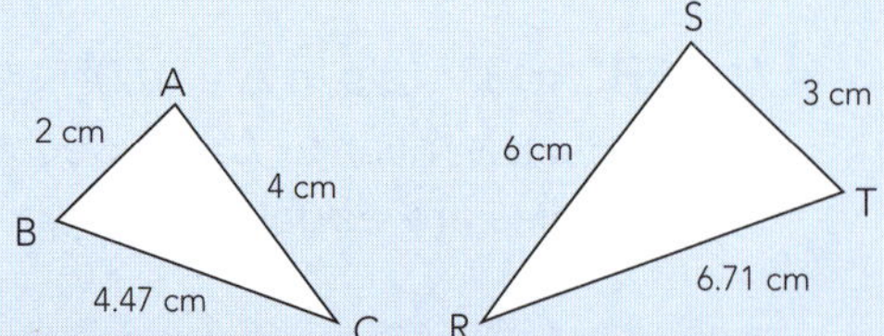

$$\frac{2}{3} = \frac{4}{6} = \frac{4.47}{6.71} = 0.67 \text{ (2 dp)}$$

$$\textbf{OR } \frac{3}{2} = \frac{6}{4} = \frac{6.71}{4.47} = 1.5 \text{ (2 dp)}$$

∴ Δ**ACB** is similar to Δ**SRT** because the sides are in proportion.

Biggest, middle, smallest.

2

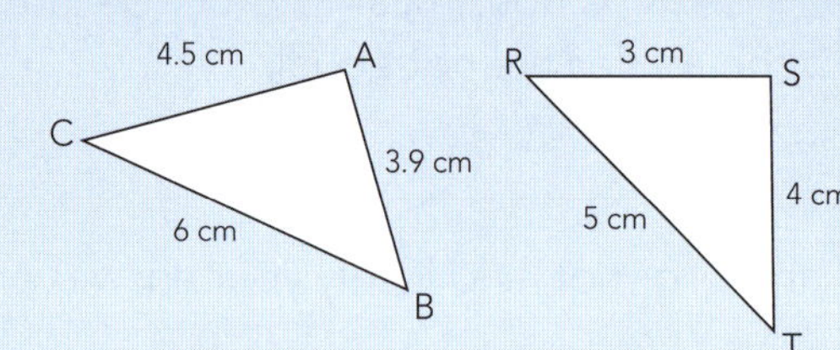

$$\frac{6}{5} = 1.2 \qquad \frac{4.5}{4} = 1.1 \qquad \frac{3.9}{3} = 1.3$$

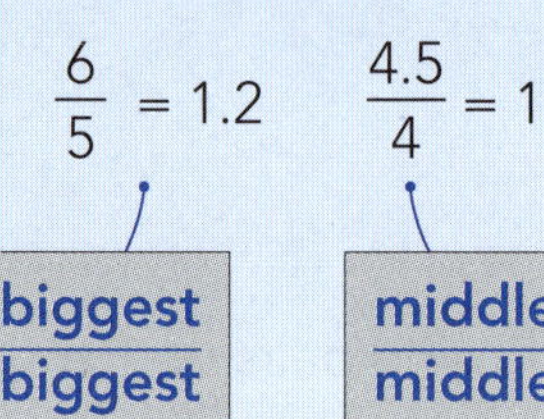

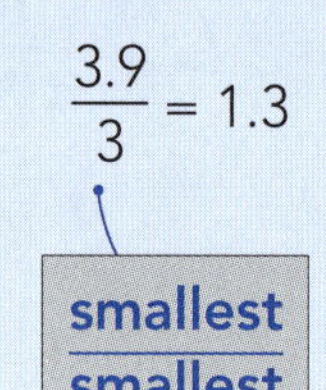

∴ Δ**ABC** is not similar to Δ**STR** because the sides are not in proportion.

State whether the following pairs of triangles are similar, and justify your answer.

1

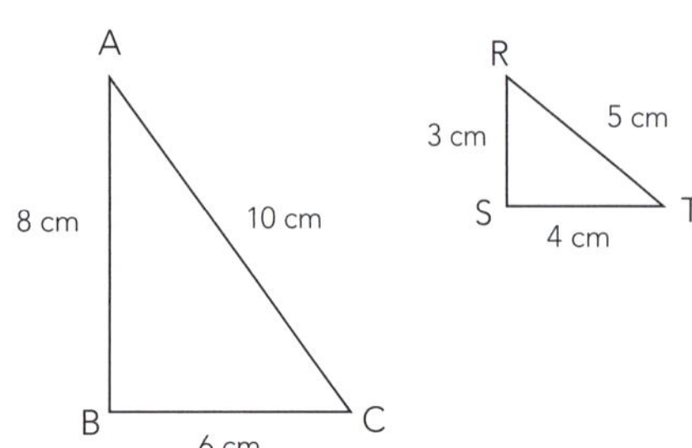

____

____

____

2

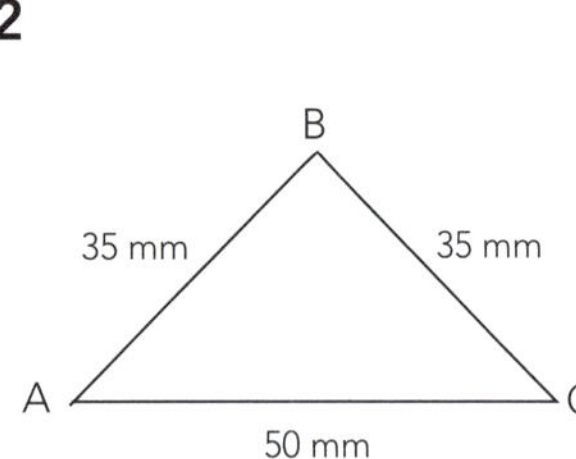

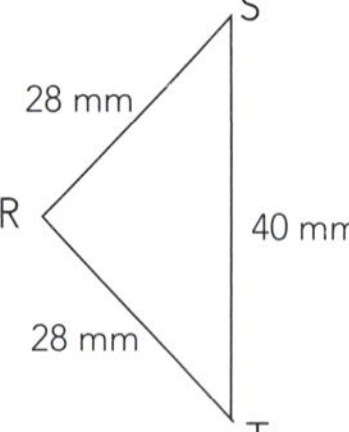

____

____

____

3

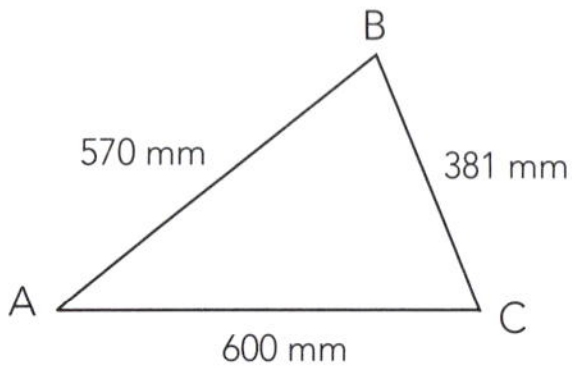

S
254 mm
395 mm
R
T
400 mm

____

____

____

ISBN: 9780170371629

Calculating unknown sides

Because the sides of similar triangles are in proportion, if we know one pair of corresponding sides (for example, both the longest sides) and we are given the length of another side (for example, a shortest side), we can calculate the length of its corresponding side.

Example: Calculate the side *y*.

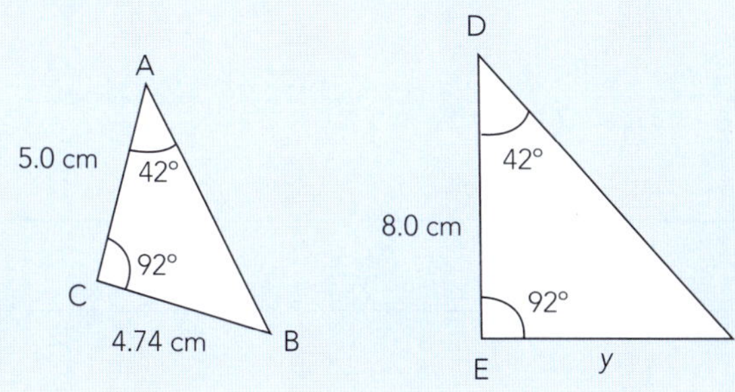

ΔABC is similar to ΔDFE ($\angle$s in Δ = 180°)

$$\therefore \frac{AC}{DE} = \frac{CB}{EF}$$

$$\frac{5.0}{8.0} = \frac{4.74}{y}$$

$$5.0y = 8.0 \times 4.74$$

$$y = 7.584 \text{ cm}$$

Calculate the missing sides in the following triangles.

1

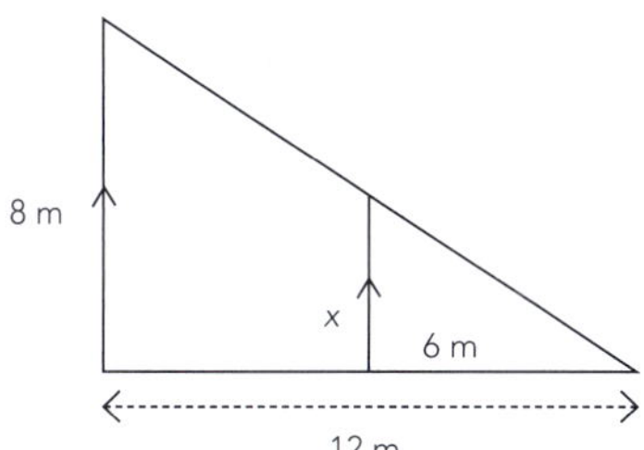

2

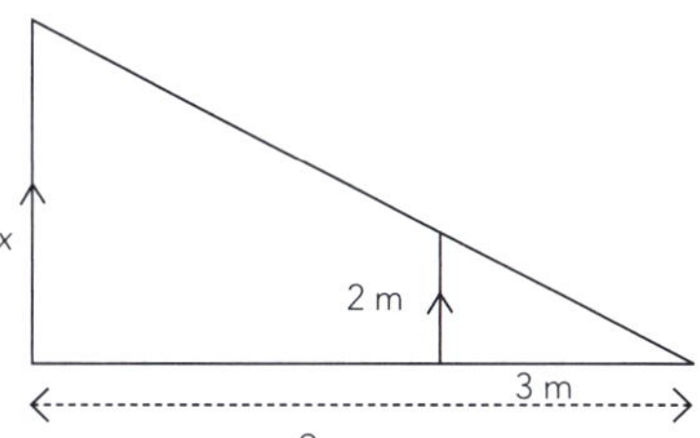

3

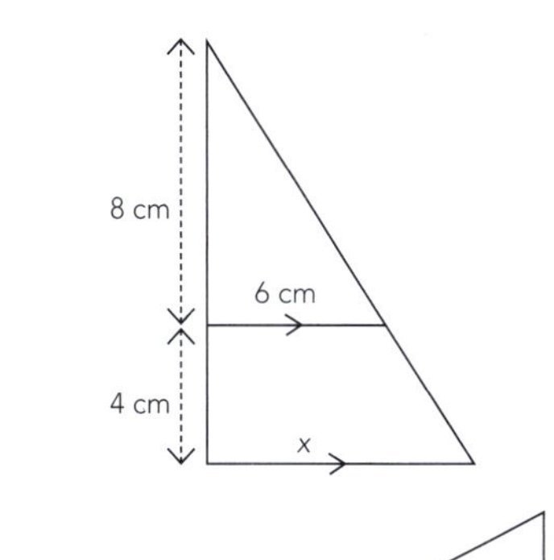

4

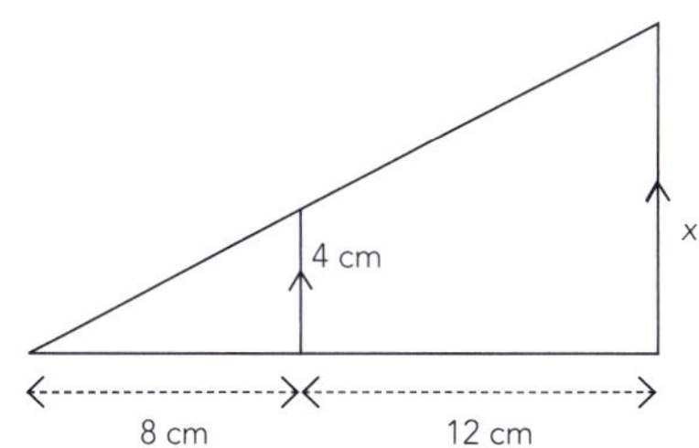

ISBN: 9780170371629

5

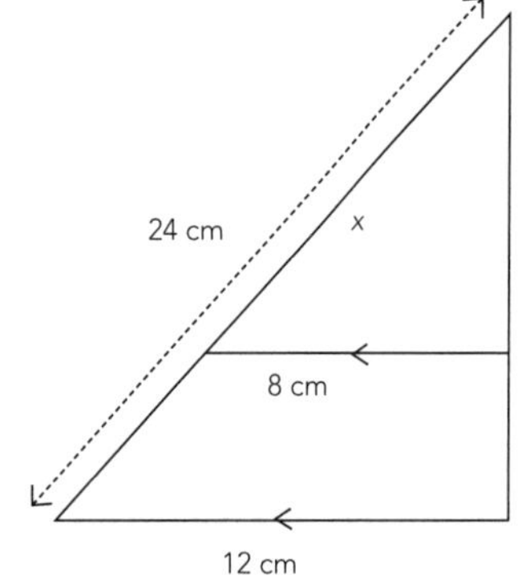

6

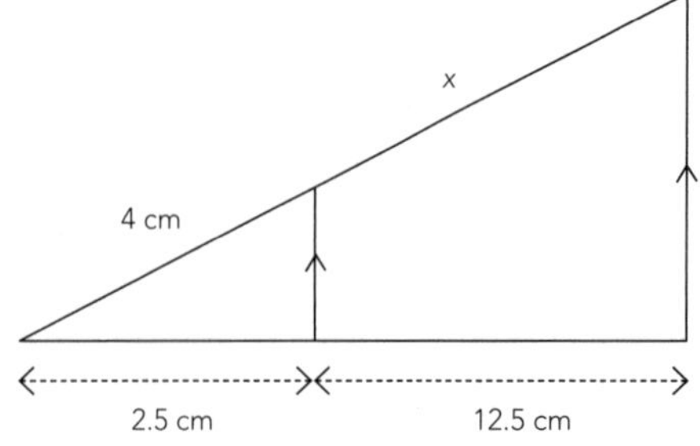

7

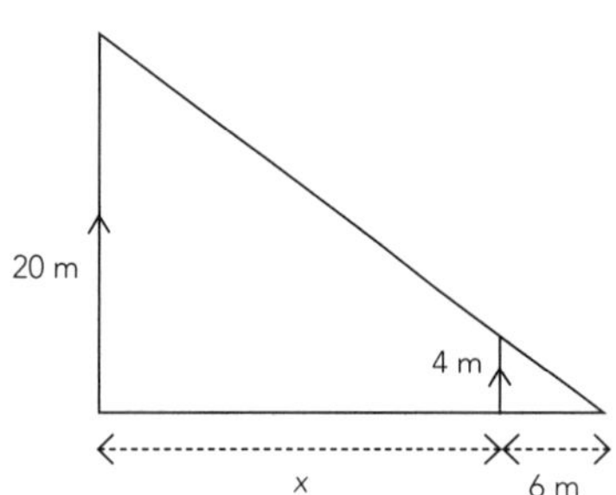

8

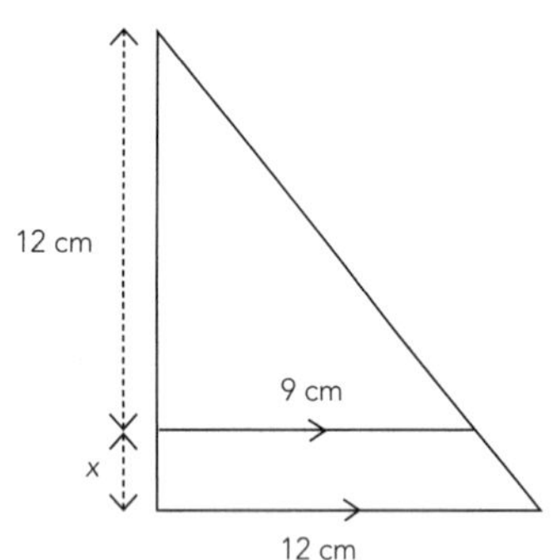

9

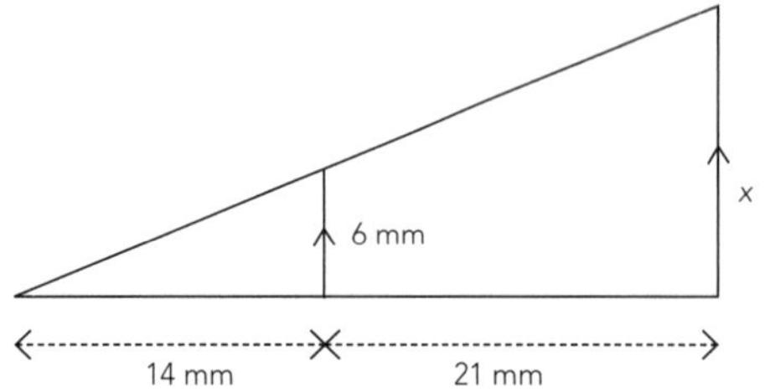

ISBN: 9780170371629

10

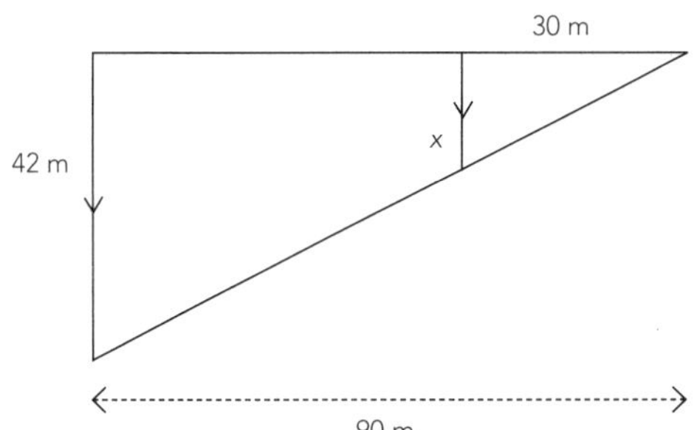

11

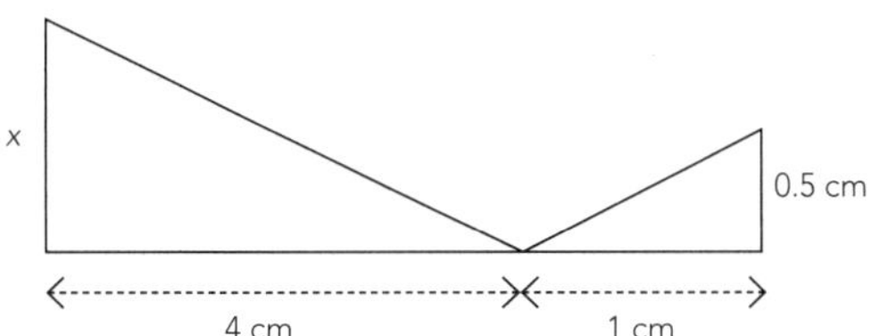

12

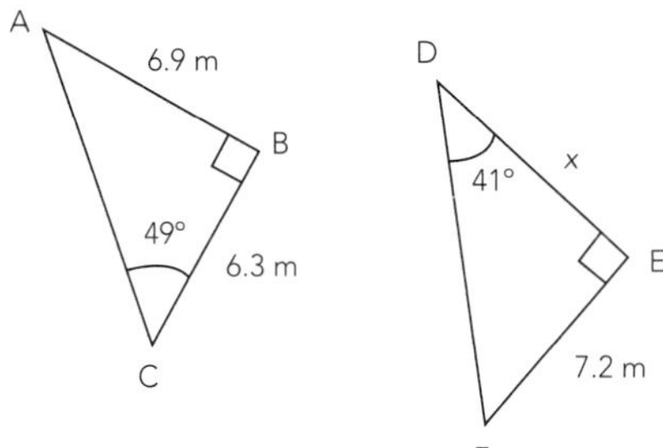

13

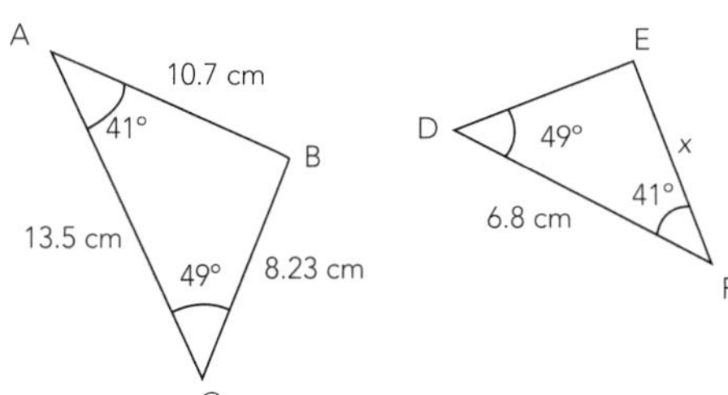

14

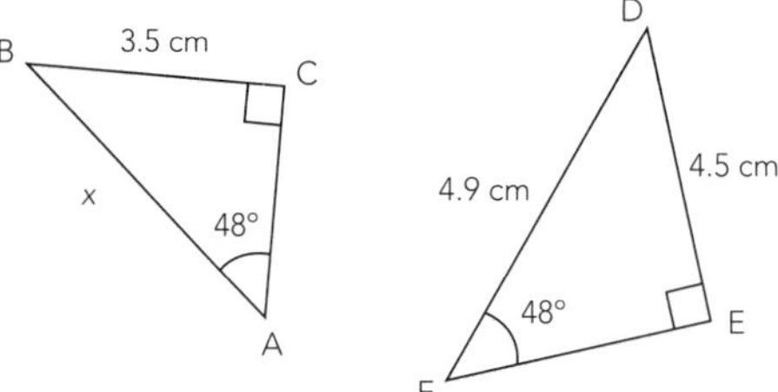

ISBN: 9780170371629

Scale diagrams

- Scale diagrams and maps are **similar shapes** to whatever they represent.
- Lengths are in proportion, so unknown lengths can be calculated.

Example one: A rectangular paddock is shown on a map:

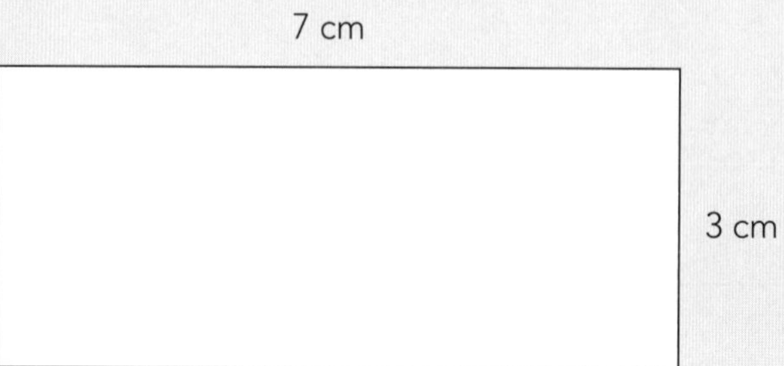

If the actual length of the short side of the paddock is 96 m, calculate the length (L) of the long side.

$$\frac{\textbf{short side on the map}}{\textbf{short side on the ground}} = \frac{\textbf{long side on the map}}{\textbf{long side on the ground}}$$

$$\frac{3}{96} = \frac{7}{L}$$

$$3L = 7 \times 96$$

$$L = \frac{7 \times 96}{3}$$

$$L = 224 \text{ m}$$

Example two: Aneke is making a scale drawing of the cross-section of a roof. The top of the roof is 1.2 m higher than the base, and the roof is 3.8 m wide.

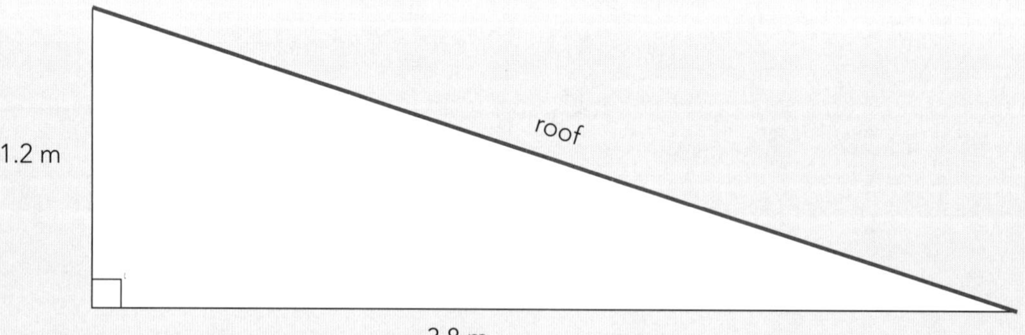

On her scale drawing, she makes the height 4 cm. How wide will she need to make the roof?

$$\frac{4}{1.2} = \frac{W}{3.8}$$

$$1.2W = 4 \times 3.8$$

$$W = \frac{4 \times 3.8}{1.2}$$

$$W = 12.67 \text{ cm}$$

ISBN: 9780170371629

Calculate the unknown lengths.

1 The diagram represents a rectangular paddock as it is shown on a map. The longer side of the paddock is 90 m. Use your measurements from the diagram to help you calculate the length of the shorter side (S) of the paddock.

S

2 A cricket pitch is 22.56 m long and 3.66 m wide. A scale diagram of it is 15 cm long. How wide (to the nearest mm) should it be?

22.56 m

3.66 m

3 The scale diagram shows two ski lifts. The real length of W is 800 m. Measure the length of W on the diagram, and use this to help you calculate:

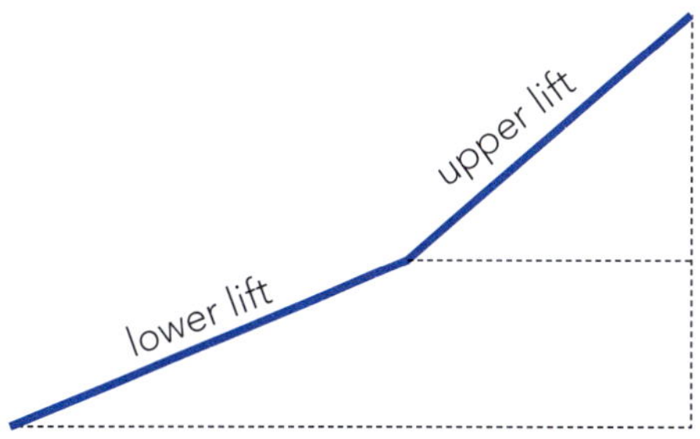

upper lift

lower lift

W

a the real length of the lower lift

b the real length of the upper lift.

4 Rua has a poster of this car. On the poster, the car is 865 mm long and 270 mm high. He wants to make a scale drawing of the car. His drawing needs to be exactly 30 mm high. How long will it be?

5 Melanie is helping to make the set for the school production. The director wants a 2.1 m high version of the soldier. Measure his height on the picture, and use this to help you calculate how wide he will be.

6 Sora is making a model plane. She needs to calculate the width of a wing. The length of the model wing will be 225 mm. Measure the length of wing on the picture, and use this to help you calculate the width on the model wing.

length

width

7 The wingspan of a monarch butterfly is 10 cm. Measure the wingspan on the scale diagram, and use it to help you calculate the length of the body of a monarch butterfly.

ISBN: 9780170371629

3D problems

- When asked to find the length of a side or an angle in a 3D shape, it is a good idea to draw the 2D figure within which the 3D shape lies.
- These usually involve right-angled triangles, so you can expect to use the Theorem of Pythagoras and trigonometry.

Example: This figure represents a cuboid.

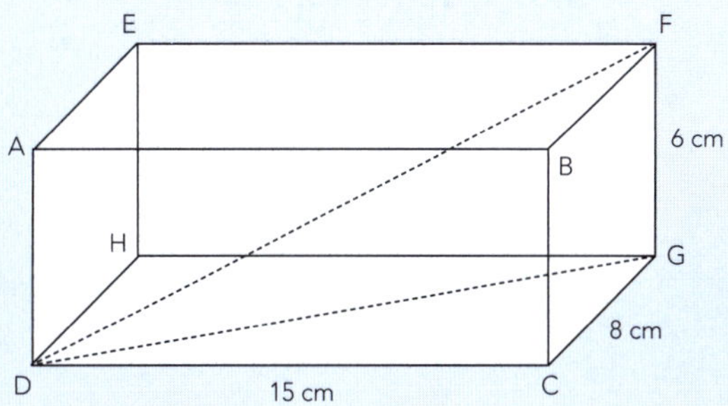

a Calculate the length of DG.
DG lies in a triangle on the 'floor' of the cuboid.

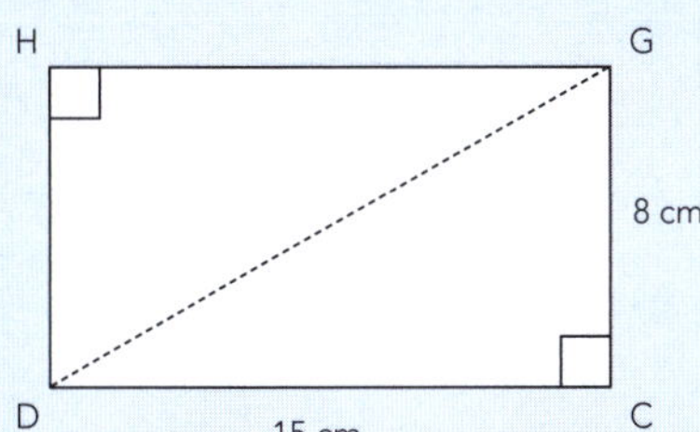

$DG^2 = 15^2 + 8^2$ (Pythagoras)
$\therefore DG = 17$ cm

b Calculate the length of DF. DF lies on the triangle DFG.

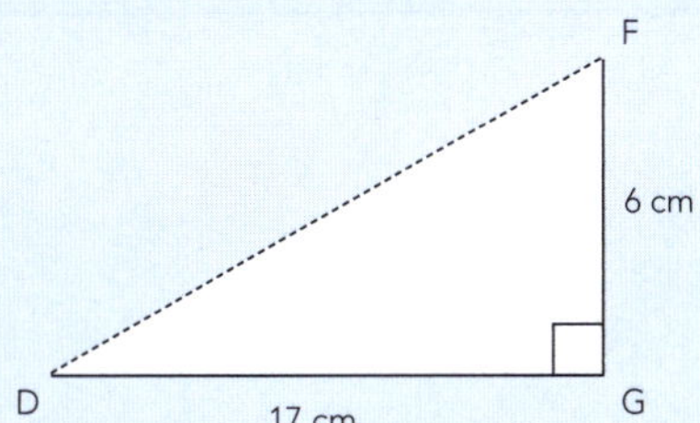

$DF^2 = 17^2 + 6^2$ (Pythagoras)
$\therefore DF = 18.03$ cm

c Calculate the angle between the line DF and the plane. ∠FDG lies on triangle DFG:

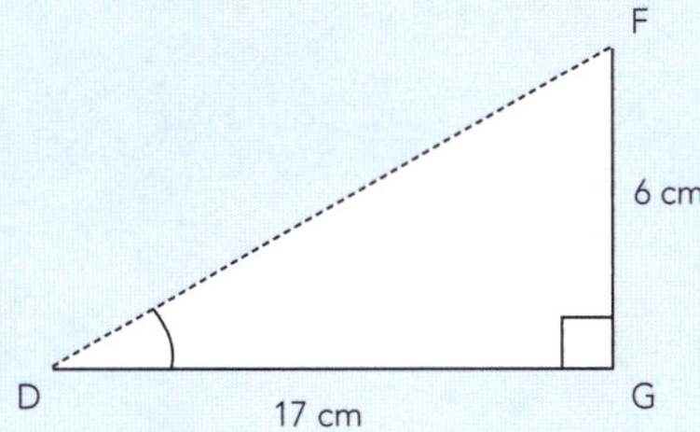

$\tan FDG = \frac{6}{17}$
$\therefore \angle FDG = 19.4°$

d Calculate the angle between the plane HFD and the plane DHGC:

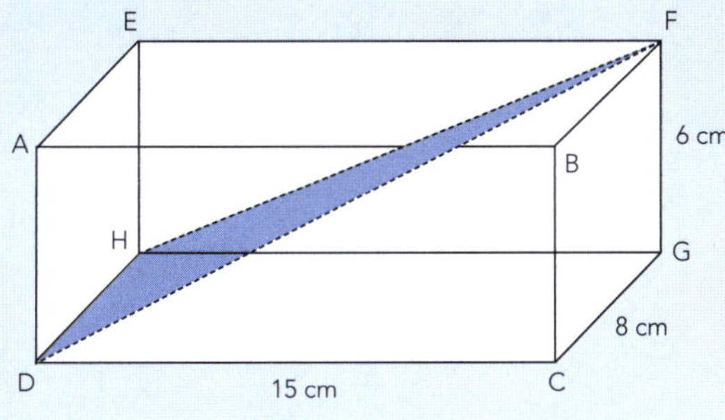

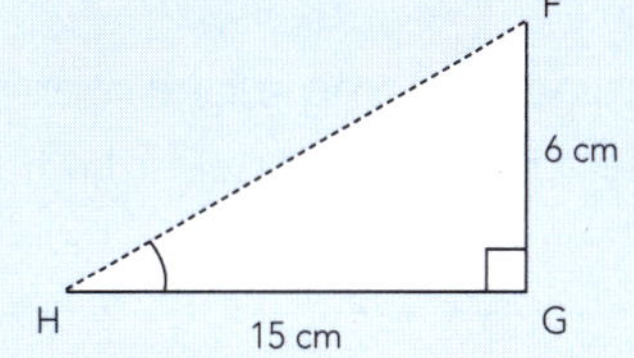

$\tan FHG = \frac{6}{15}$
$\therefore \angle FHG = 21.8°$

ISBN: 9780170371629

Calculate the missing angles and sides for the following.

1 This figure shows a cuboid (box).

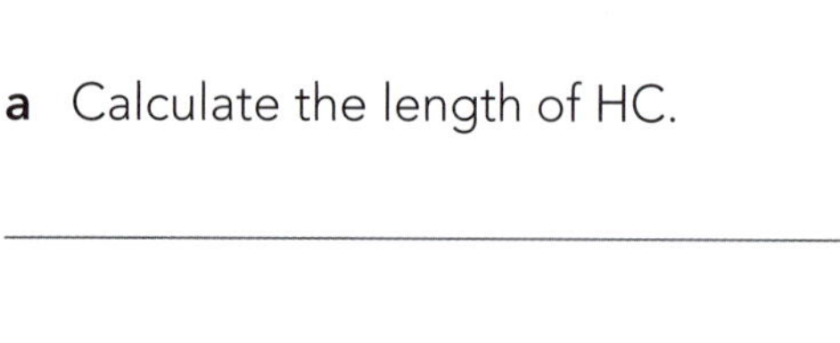

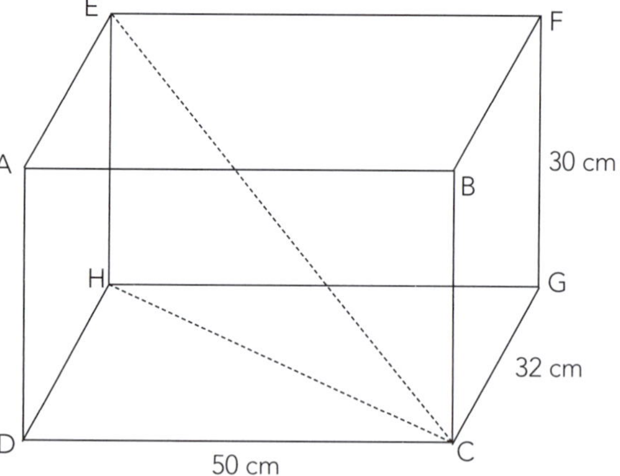

a Calculate the length of HC.

b Calculate the length of EC.

c Calculate the angle between the line EC and the plane HGCD.

d Calculate the angle between the plane ECH and plane EFGH.

e Calculate the angle between the plane EHC and plane AEHD.

ISBN: 9780170371629

2 A vertical pole (TA) on a flat roof is being used to support two 6 m long strings of Christmas lights (TB and TD). The bottom ends of the strings of lights (BD) are 6 m apart, and the same distance from A. A third string of lights will be placed from T to a point halfway between B and D.

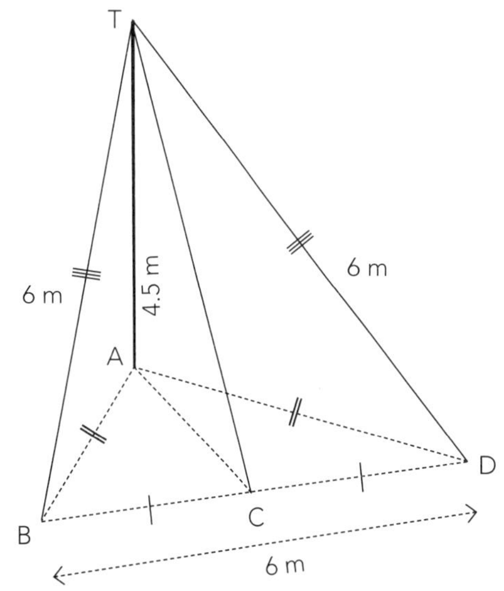

a Calculate the distance between the base of the pole (A) and the end of the 6 m string (D).

b Calculate the angle between the 6 m string of lights (TD) and the roof.

c Calculate the distance from A to the bottom of the middle string of lights.

d How long is TC, the middle string of lights?

e Calculate the angle between TC and the roof.

ISBN: 9780170371629

3 An 8 m high pyramid is built on a square base with sides that are 8 m long.
The top (T) of the pyramid is 8 m above the centre of the square base.
Some answers will need several steps of working.

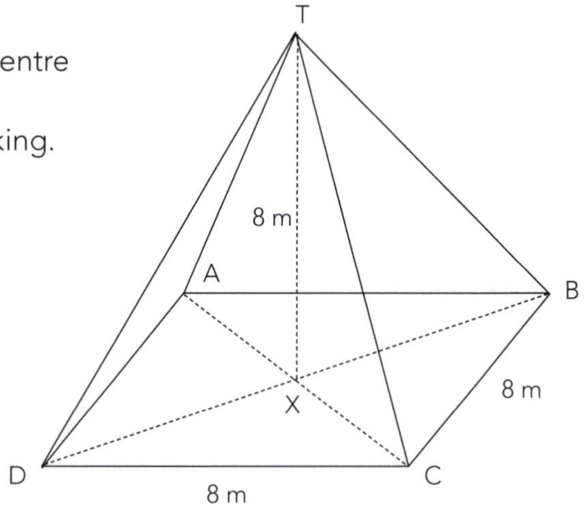

a Calculate the length of the edge TB.

b Calculate the angle between the edge (TB) and the base.

c Calculate the angle between the face TCB and the base of the pyramid.

d A second identical pyramid is built next to the original one. Calculate the size of angle TCS.

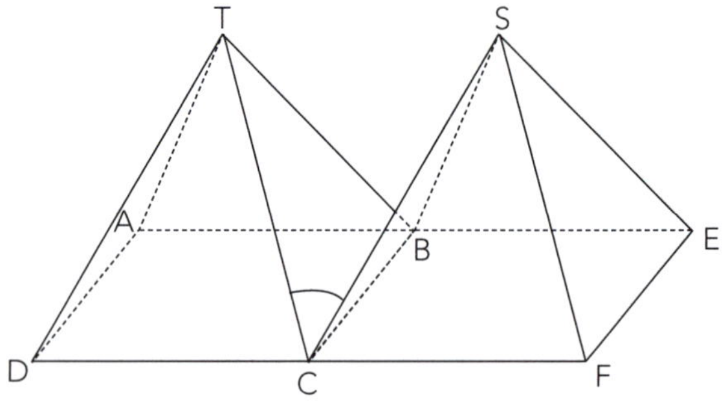

 ISBN: 9780170371629

Practice tasks

Practice task one

Jenna has based the blue design on the black dashed tessellation. All the arms of the black figures are 2 cm wide. All the angles in the black figures are right angles. The corners of one section of the blue pattern have been labelled to help you.

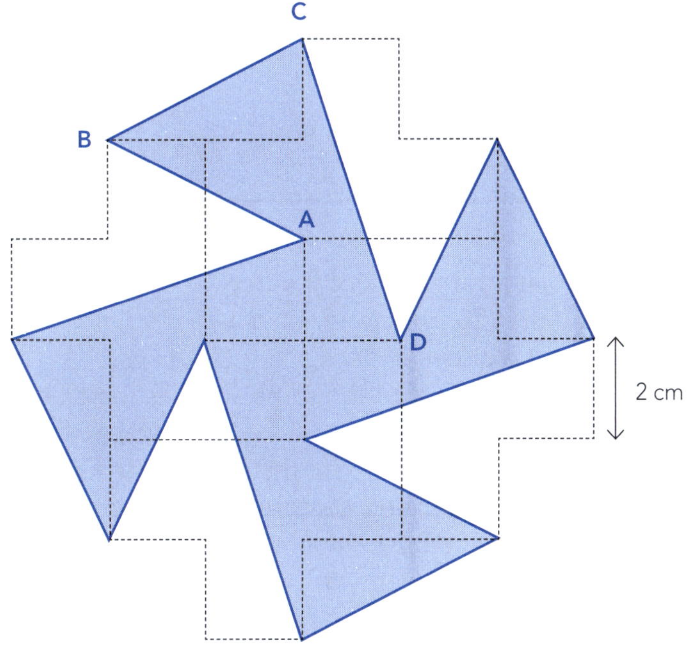

- Calculate the perimeter of the blue shape.
- Jenna's friend suggests that she could make the blue shape bigger and more interesting by adding a right-angled isosceles triangle to each arm, with BC forming the hypotenuse. Calculate the perimeter of this new shape.

Practice task two

Ali needs to know the height of the flagpole in his school grounds. He measures the length of the shadow cast by a tree, and the length of the shadow cast by the flagpole. He then measures the height of the tree. The diagram shows a scale drawing of the tree and its shadow.

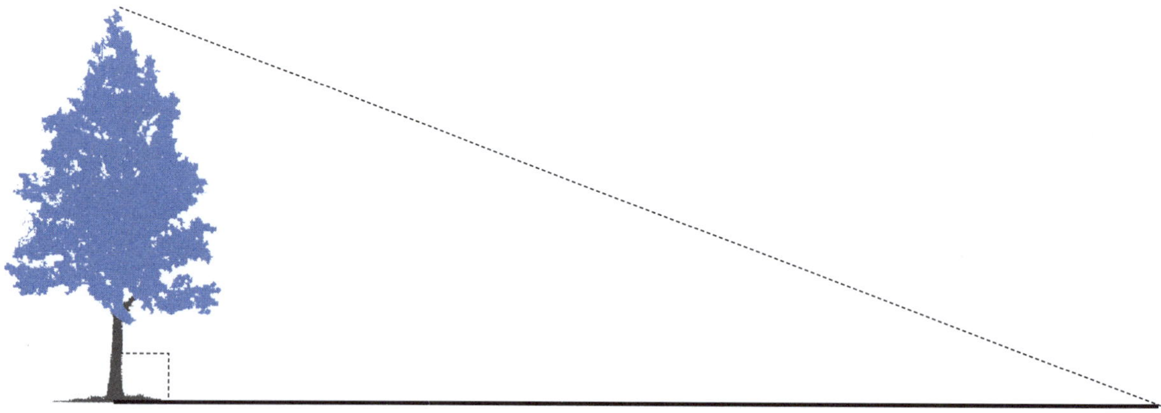

- Use the scale diagram to calculate the height of the tree.
- The flagpole casts a shadow that is 23.34 m long. Calculate the height of the flagpole. Give your answer in metres to one decimal place.
- If the flagpole and its shadow were drawn to the same scale as the diagram above, how high and wide would they be on the diagram?
- The caretaker wants to take the flagpole down for restoration. He can work on it outside his workshop during the day, but he will need to store it inside the workshop at night. His workshop is 3.8 m wide, 6.9 m long and 3.2 m high. Will the flagpole fit diagonally inside the workshop at night? Support your answer with calculations.

ISBN: 9780170371629

Practice task three

Esther has been dropped off by her bus at A, and is walking to meet a friend at C. The grey rectangles represent parks. The roads around the parks are 10 m wide.

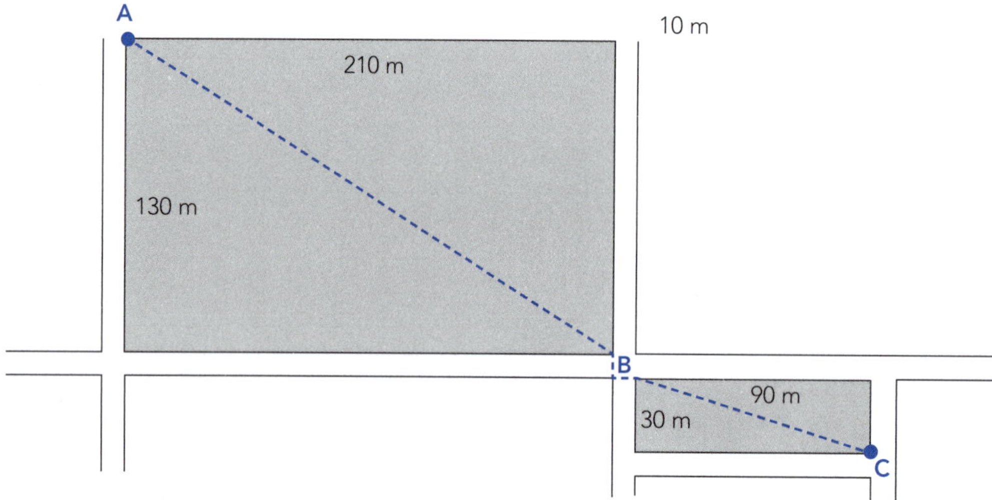

- From A she can walk diagonally across the first park to B, cross the two pedestrian crossings at B, and then cross the second park diagonally to C.
  Alternatively, she can walk on the footpaths around the edge of the first park, across the crossings at B and then on the footpaths around the edge of the second park to point C. How much shorter is her walk from A to C if she cuts diagonally across both parks?

A fiesta is to be held in the smaller park. There is a 3 m high pole in each corner. Four strings of lights will be stretched tightly from each corner pole to a fifth pole in the centre of the park.

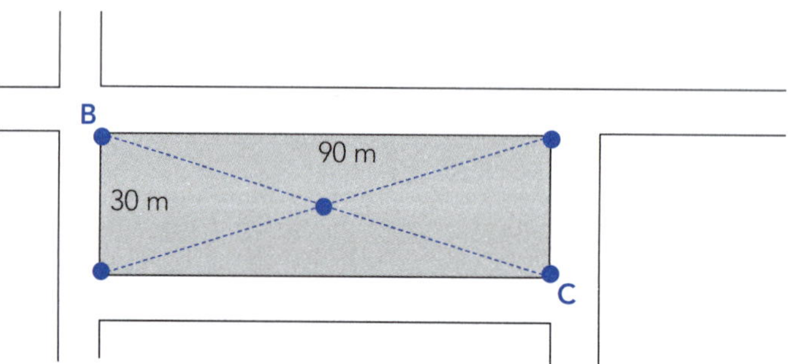

The strings of lights must make an angle of 3° with the horizontal line between the tops of the 3 m poles. A cross-section through BC is shown below.

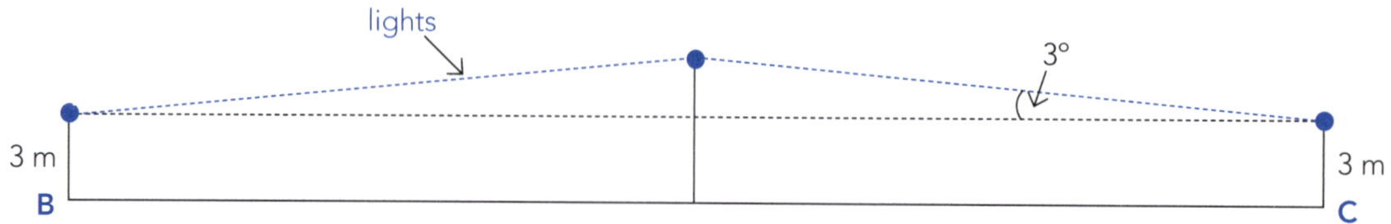

- Calculate the required height of the central pole.
- Calculate the length of each string of lights.

 ISBN: 9780170371629

Practice task four

Laura and Chris are making origami flowers, which start with a piece of paper that must be a regular pentagon (all sides the same). The first folds are shown as dotted lines on the diagram. The diagram is drawn to scale.

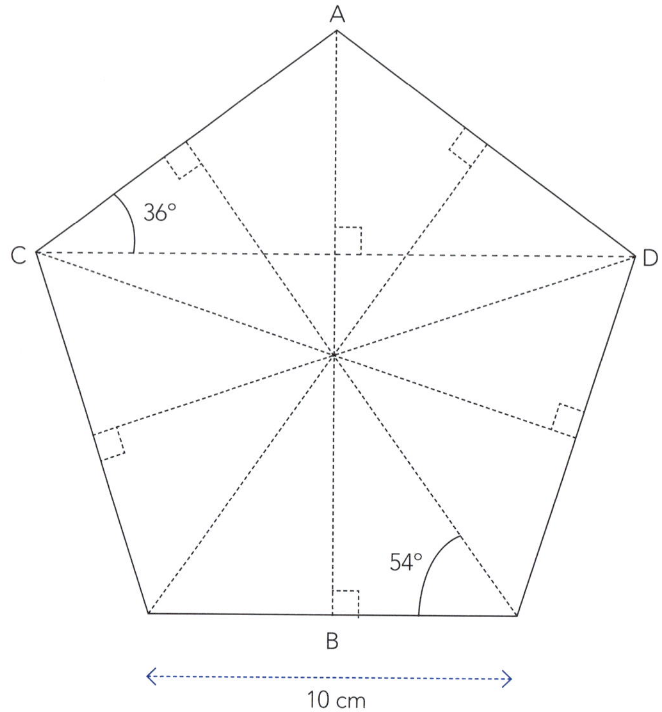

Chris maintains that each piece of paper will need to be square (so AB = CD). Laura doesn't think so.

- Measure the length of AB on the diagram, and use the scale to calculate the actual height of AB.
- Use trigonometry to calculate the actual height AB. Do your two answers agree?
- Measure the width of CD on the diagram, and use the scale to calculate the actual width of CD.
- Use trigonometry to calculate the actual width CD. Do your two answers agree?
- Is Laura or Chris correct?

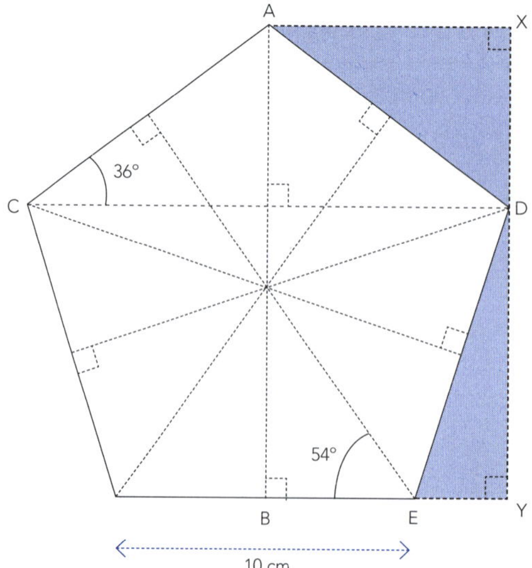

ISBN: 9780170371629

They are going to use the light blue triangles for another project.

- Calculate the dimensions of these triangles.

Answers

Lengths and areas are rounded to a maximum of 4 sf.
Angles are rounded to a maximum of 1 dp.
Professional judgement should apply.

Metric units (pp. 6–9)

Length (p. 6)

1	270 mm	**2**	800 cm
3	2000 m	**4**	90 mm
5	9.5 cm	**6**	1.4 km
7	1700 cm	**8**	8.2 cm
9	170 cm	**10**	113 mm
11	6.85 m	**12**	2380 m
13	700 m	**14**	3.72 m
15	2640 m	**16**	815 cm
17	7.350 km	**18**	560 cm
19	40 m	**20**	0.06 m

Area (p. 7)

1	20 000 cm^2	**2**	1.7 cm^2
3	400 mm^2	**4**	15.46 cm^2
5	12 000 cm^2	**6**	0.6932 ha
7	25 600 m^2	**8**	800 m^2
9	6040 m^2	**10**	1.275 ha
11	7.65 cm^2	**12**	0.15 ha
13	4.284 m^2	**14**	40 m

Sensible units (p. 8)

1	cm	**2**	cm or m
3	m	**4**	ha
5	mm^2	**6**	km
7	m^2	**8**	m
9	mm or cm	**10**	cm^2
11	mm	**12**	cm
13	m	**14**	mm or cm
15	m or cm	**16**	m^2
17	cm^2	**18**	ha

Estimating lengths and areas (p. 9)

1 The length of your calculator would most likely be 15 cm.
2 The circumference of your head is about 55 cm.
3 The average length of the South Island is about 840 km.
4 The diameter of a pencil lead would be about 2 mm.
5 A length of your classroom would most likely be 8 m.
6 The length of a rugby field is about 100 m.
7 The area of a rugby field is about 0.5 ha.
8 The area of the front cover of this book is about 600 cm^2.
9 The area of your bedroom would most likely be 9 m^2.
10 The length of your school hall would most likely be 30 m.
11 The surface area of your skin would most likely be 1.5 m^2.
12 The average height of a Year 9 student would be about 1.5 m.
13 The surface area of a potato would be about 120 cm^2.
14 The area of your school grounds would most likely be 8 ha.
15 The area of your section at home would most likely be 0.05 ha.
16 The area of the whiteboard in your classroom would most likely be 3.5 m^2.

Theorem of Pythagoras (pp. 10–18)

Finding the length of the hypotenuse (pp. 10–12)

1	14.87 m	**2**	5 m
3	16.55 cm	**4**	15 cm
5	8.062 km	**6**	6.940 m
7	20 m	**8**	3.027 km or 3027 m
9	511.2 m	**10**	42.43 mm
11	0.2802 km	**12**	62.77 cm
13	57.69 cm or 0.5769 m	**14**	46.14 m
15	63.64 cm	**16**	30.10 cm

Finding the lengths of short sides (pp. 13–14)

1	9 m	**2**	2.646 cm
3	9.798 cm	**4**	7.937 cm
5	5.745 km	**6**	3.814 m
7	2.973 km	**8**	9.165 m
9	393.2 m	**10**	40 mm
11	168.8 mm	**12**	2.664 m
13	3.862 km	**14**	6.928 m

ISBN: 9780170371629

Mixing it up (pp. 15–16)

1	33.97 m	**2**	20.17 cm
3	9.520 cm	**4**	0.3982 km
5	8.386 m	**6**	312.0 mm
7	2.993 km	**8**	206.2 cm
9	4.649 km	**10**	49.50 cm
11	15.97 cm	**12**	6.125 m
13	87.46 cm	**14**	67.27 m
15	3.991 m	**16**	12.69 m

Applications (pp. 17–18)

1	1.7 m	**2**	3.747 m
3	8.544 cm	**4**	21.02 cm

5 AB = 14.14 cm, so perimeter = 48.28 cm

6 $2(AD)^2 = 289 \rightarrow AD = 12.02$ cm

Trigonometry (pp. 19–43)

1

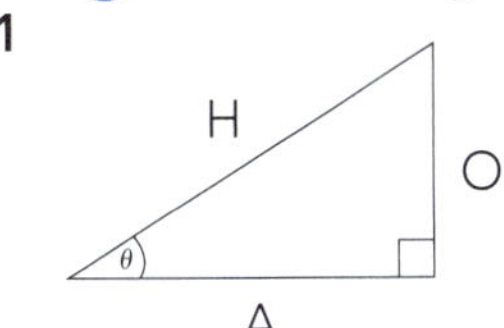

2

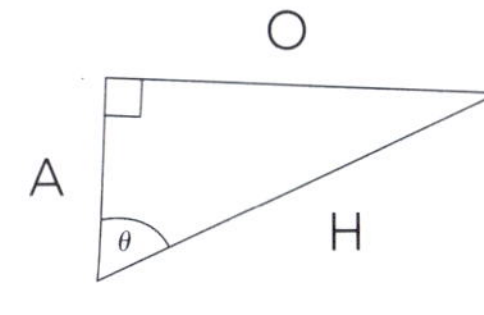

3

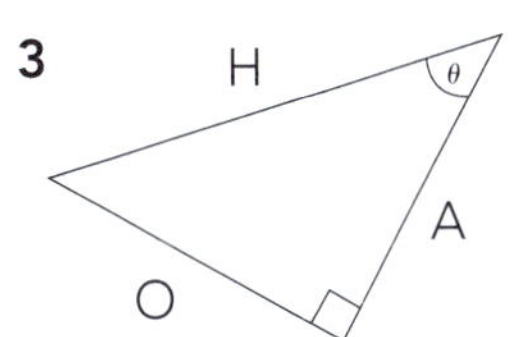

4

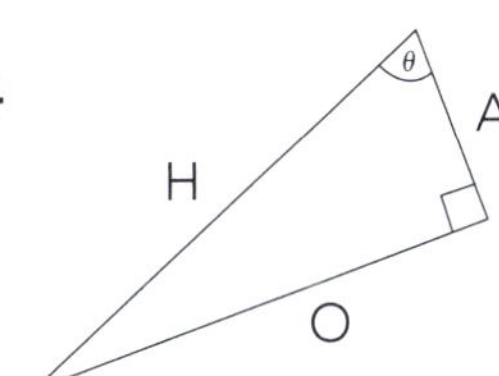

5

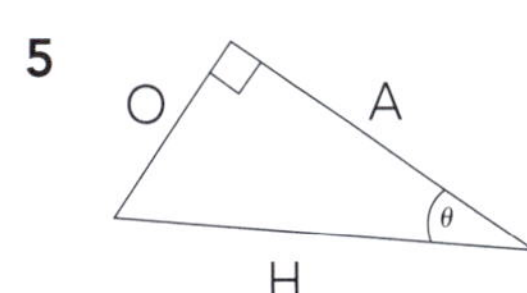

6

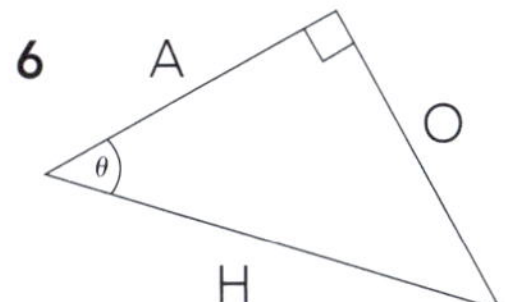

Finding sides using the sine (pp. 20–23)

1	8.632 m	**2**	6.223 cm
3	2.868 cm	**4**	5.416 cm
5	8.074 km	**6**	4.800 m
7	0.0726 km	**8**	12.94 m
9	94.95 mm	**10**	42.43 mm
11	49.93 cm	**12**	1.907 km
13	23.11 cm	**14**	134.2 mm
15	15.56 cm	**16**	139.5 cm
17	0.7873 m	**18**	2.782 km
19	65.05 mm	**20**	4.764 km

Finding sides using the cosine (pp. 24–26)

1	13.99 m	**2**	8.030 cm
3	8.254 cm	**4**	0.7461 km
5	2.743 km	**6**	86.09 cm
7	4.177 km	**8**	13.07 cm
9	29.30 cm	**10**	551.5 mm
11	0.8323 km	**12**	8.537 cm
13	46.71 cm	**14**	771.6 mm
15	4.620 m	**16**	77.10 cm

Finding sides using the tangent (pp. 27–29)

1	8.719 m	**2**	19.27 cm
3	9.602 cm	**4**	28.17 cm
5	0.4442 km	**6**	0.9980 m
7	1.033 km	**8**	29.21 m
9	717.1 m	**10**	61.58 mm
11	55.35 cm	**12**	12.52 m
13	3.094 km	**14**	1548 mm
15	76.47 cm	**16**	54 mm

Finding sides — mixing it up (pp. 30–31)

1	15 m	**2**	2.828 cm
3	2.347 cm	**4**	13.34 cm
5	5.978 km	**6**	3.365 m
7	0.7258 km	**8**	30.71 m
9	115.5 cm	**10**	0.5402 km
11	14.02 km	**12**	39.59 cm
13	$x = 67.62$ cm $y = 40.69$ cm	**14**	$x = 2.809$ km $y = 4.495$ km
15	$x = 11.84$ m $y = 14.63$ m	**16**	$x = 24.27$ cm $y = 11.84$ cm

Finding sides — applications (pp. 32–33)

1 1.329 m

2 Belt is 7.572 m.
Horizontal distance is 6.862 m.

3 Starts 3.334 m from classroom.
Length = 3.346 m

4 $h = 2.498$ m
Roof needs to be 6.499 m long.

5 3.788 m + 1.52 m = 5.308 m

6 $h = 504.1$ m

Finding angles using the sine (pp. 34–35)

1	61.9°	**2**	33.2°
3	37.6°	**4**	30.8°
5	50.9°	**6**	36.5°
7	52.3°	**8**	31.6°
9	66.2°	**10**	26.5°
11	37.7°	**12**	9.5°

Finding angles using the cosine (pp. 36–37)

1	28.1°	**2**	54.8°
3	49.2°	**4**	57.5°
5	27.7°	**6**	46.5°
7	44.9°	**8**	59.6°
9	42.1°	**10**	57.7°
11	34.5°	**12**	53.6°

ISBN: 9780170371629

Finding angles using the tangent (pp. 38–39)

1 48.6°
2 59.0°
3 27.2°
4 59.3°
5 40.5°
6 52.1°
7 41.2°
8 59.6°
9 44.2°
10 74.7°
11 48.0°
12 53.4°

Finding angles — mixing it up (pp. 40–41)

1 34.3°
2 62.9°
3 25.2°
4 41.4°
5 56.6°
6 42.9°
7 60.7°
8 45.0°
9 32.9°
10 40.0°
11 41.0°
12 76.2°
13 64.5°
14 52.3°
15 51.5°
16 61.1°

Finding angles — applications (pp. 42–43)

1 64.4°
2 33.7°
3 Angle made by sun's rays and ground = 49°.
Height of tree = 6.17 m.
4 33.7°
5 **Either:** $\sin^{-1}(3 \div 15) = 11.5°$ ∴ yes because the angle would be less than 12° maximum angle.
Or: minimum travellator length required is 3 ÷ sin 12° = 14.43 m. This is less than 15 m travellator, so it will comply.
Or: Max angle ⇒ height = 15 sin 12° = 3.12 m. ∴ travellator will comply.
Or: maximum height allowed = 15 x sin12° = 3.119 m. This is more than the height required, so it will comply.
6 **Either:** $\tan^{-1}(0.34 \div 4) = 4.858°$ ∴ no because the angle would be more than 4.8°, and too steep.
Or: distance required is 0.34 ÷ tan 4.8° = 4.049 m. This is more than 4 m so the ramp won't fit.
Or: Max angle ⇒ height = 4 sin 4.8 = 0.3 m. Height required is 0.34 m in, so the ramp won't produce enough height.
Or: maximum height the ramp can be = 4 x tan 4.8 = 0.3359 m = 33.59 cm, so a ramp which complies will not be high enough.

Putting it together — with Pythagoras (pp. 44–47)

1 31.6°
2 9.793 cm
3 34.44 mm
4 49.1°
5 5.338 km
6 1.426 km
7 199.6 mm
8 54.6°
9 35.2°
10 **a** 4.8°
b 12.04 units
11 62.8°
12 **a** 5.492 cm
b 56.5°
13 **a** 9.970 cm
b 7.962 cm
14 **a** 26.6°
b 8.94 cm
15 ∠BAC = 48°
AD = 3.546 cm
DC = 8.019 cm
AC = 11.57 cm
16 ∠BYX = 45° (isos Δ)
h = 25 cm
17 AB = 7.52 cm
BC = 4 cm
∴ AC = 11.52 cm
18 AC = 10 cm
AD = 12.81 cm
AE = 16.25 cm
Perimeter = 48.25 cm

Similar triangles (pp. 48–53)

Justification using angles (p. 49)

1 ∠CAB = ∠EDF = 42°
∴ ΔABC is similar to ΔDFE
2 ∠ABC = 51° and ∠FDE = 39°
∴ ΔABC is similar to ΔEFD
3 ∠BAC = 38° and ∠EDF = 53°
∴ ΔABC is not similar to ΔFED
4 ∠ABC = 48° and ∠FED = 42°
∴ ΔABC is similar to ΔEFD
5 ∠AEB = 90° (corr ∠s, // lines)

ISBN: 9780170371629

∠BAE is common to both Δs

∴ ΔAEB is similar to ΔADC

Justification using sides (p. 50)

1 $\frac{10}{5} = \frac{8}{4} = \frac{6}{3} = 2$

∴ ΔABC is similar to ΔTSR

2 $\frac{50}{40} = \frac{35}{28} = \frac{35}{28} = 1.25$

∴ ΔABC is similar to ΔTRS

3 $\frac{600}{400} = 1.5 \quad \frac{381}{254} = 1.5 \quad \frac{570}{395} = 1.44$

∴ ΔABC is not similar to ΔTSR

Calculating unknown sides (pp. 51–53)

1 $x = 4$ m
2 $x = 6$ m
3 $x = 9$ cm
4 $x = 10$ cm
5 $x = 16$ cm
6 $x = 20$ cm
7 $x = 24$ m
8 $x = 4$ cm
9 $x = 15$ mm
10 $x = 14$ m
11 $x = 2$ cm
12 $\frac{x}{6.9} = \frac{7.2}{6.3} \rightarrow x = 7.886$ m
13 $\frac{x}{10.7} = \frac{6.8}{13.5} \rightarrow x = 5.390$ cm
14 $\frac{x}{4.9} = \frac{3.5}{4.5} \rightarrow x = 3.8\dot{1}$ cm

Scale diagrams (pp. 54–56)

Your answers may differ a little because rulers vary. Check with your teacher if your answer is different.

1 $S = 75$ m
2 $W = 2.4$ cm
3 a lower lift = 506.7 m
b upper lift = 440 m
4 length = 96.11 mm
5 width = 61.25 cm
6 5.4 cm
7 4.17 cm

3D problems (pp. 57–60)

1 a HC = 59.36 cm
b EC = 66.51 cm
c ∠ECH = 26.8°
d ∠CHG = 32.6°
e ∠DHC = 57.4°

2 a AD = 3.969 m
b ∠TDA = 48.6°
c AC = 2.599 m
d TC = 5.197 m
e ∠TCA = 60.0°

3 a TB = 9.798 m
b ∠TBD = 54.7°
c Angle = 63.4°
d Angle = 180° – 2(65.9°) = 48.2°

Practice tasks (pp. 61–67)

Practice task one (p. 61)
Original shape:

Length AB = $\sqrt{2^2 + 4^2}$ = 4.472 cm
Length BC = $\sqrt{2^2 + 4^2}$ = 4.472 cm
Length CD = $\sqrt{2^2 + 6^2}$ = 6.325 cm

Perimeter = 4(2(4.472) + 6.325)
= 61.08 cm

New shape:
Triangle side = 4.472 sin 45°
= 3.162 cm

New perimeter = 4(4.472 + 2(3.162) + 6.325
= 68.48 cm

Practice task two (pp. 62–63)
Your answers may differ a little due to rounding. Check with your teacher if your answer is different.

Height of the tree: $\frac{5.7}{14.25} = \frac{x}{5.25}$
$x = 2.1$ m

Height of the flagpole: $\frac{23.34}{5.7} = \frac{h}{2.1}$
$h = 8.6$ m

Scale on diagram is 1 m = 2.5 cm

Height of the diagram = 8.6 x 2.5 = 21.5 cm
Width of diagram = 23.34 x 2.5 = 58.35 cm

Length that will fit in the workshop
= $\sqrt{(\sqrt{3.82^2 + 6.9^2})^2 + 3.2^2}$
= 8.502 m

So the flagpole will not fit inside the workshop.

Practice task three (pp. 64–65)
Walking diagonally:

D = $\sqrt{130^2 + 210^2} + 10 + 10\sqrt{90^2 + 30^2}$
= 361.85 m

Walking around the edge:

$D = 130 + 210 + 10 + 10 + 90 + 30 = 480$ m

$\therefore$ Difference = 118.15 m

Height of the central pole:

$$\begin{aligned}\text{Height} &= 3 + 47.435 \tan 3^\circ \\ &= 5.49 \text{ m}\end{aligned}$$

Length of each string of lights:

$$\begin{aligned}\text{Length} &= \frac{47.435}{\cos 3^\circ} \\ &= 47.5 \text{ m}\end{aligned}$$

Practice task four (pp. 66–67)

Height $AB = 5 \tan 54^\circ + \frac{5}{\cos 54^\circ} = 15.39$ cm

Length $CD = 2(10 \cos 36^\circ) = 16.18$ cm

$\therefore$ Laura is correct because $15.39 \neq 16.18$

Triangle AXD:

$AX = 10 \cos 36^\circ = 8.090$ cm

$XD = 10 \sin 36^\circ = 5.878$ cm

$AD = 10$ cm

Triangle DEY:

$DY = 10 \sin 72^\circ = 9.511$ cm

$YE = 10 \cos 72^\circ = 3.090$ cm

$ED = 10$ cm

ISBN: 9780170371629